T0331001

Managing Safety in the Drone Industry

The drone industry is one of the most exciting and dynamic sectors in the modern world, growing and developing at an exponential pace. With an increase in the usage numbers and technological sophistication advances of these devices, their future development and application are only limited by our imagination. *Managing Safety in the Drone Industry: A Practical Guide* has been written with the aim of helping those involved in drone operations develop safety structures and practices capable of tackling the increased risk of accidents and incidents as more drones of various sizes begin to fill our skies.

This book is designed to suggest practical guidelines for managing safety primarily for the new drone industry, particularly those of a safety critical nature. Based upon the author's 50 years of experience in the Aviation Industry, this book uses tried-and-tested best practices and current aviation principles adapted to drones and attempts to predict the trends of evolution which are believed to become mandatory regulations in the future. Real-life case studies are aligned with the text. Written in a style that adopts short and succinct chapters, this guide will allow the reader a thorough overview of drone safety and the future trends that the industry will face.

This book is an ideal read for any professional working in a safety critical industry needing practical guidance on drones or manned aviation. It will particularly appeal to those in aviation, health and safety, logistics, business, and management and in any industry using or considering the use of drone technology for their operations.

FIGURE 0.1 Photograph of a DJI Matrice 3D deployed from a docking station.

Managing Safety in the Drone Industry

A Practical Guide

Geoff Bain and Mark Blaney

CRC Press is an imprint of the
Taylor & Francis Group, an **informa** business

Designed cover image: A.Film (Shutterstock ID: 383122651)

First edition published 2025
by CRC Press
2385 NW Executive Center Drive, Suite 320, Boca Raton FL 33431

and by CRC Press
4 Park Square, Milton Park, Abingdon, Oxon, OX14 4RN

CRC Press is an imprint of Taylor & Francis Group, LLC

ISBN: 978-1-032-61146-4 (hbk)
ISBN: 978-1-032-62020-6 (pbk)
ISBN: 978-1-032-62022-0 (ebk)

DOI: 10.1201/9781032620220

Typeset in Times LT Std
by Apex CoVantage, LLC

Contents

PART TWO Human Factors (HF), Crew Resource Management (CRM), Team Resource Management (TRM)

PART SIX Practical Application in the RPAS (Drone) Industry

Preface

The drone industry is one of the most exciting and dynamic sectors in the modern world, growing and developing at an exponential pace. Drones are now a part of all our lives and expected to assume an even greater importance in the coming years. As numbers grow and technological sophistication advances, the future development and application is only limited by our imagination. With this development, there will logically be an increased risk of accidents and incidents as more drones ranging from nanoscale to large machines, potentially the same size as currently manned aircraft, begin to fill our skies. An obvious consequence is the need for much more stringent legal regulation of operational standards and practices. These are predicted to evolve into closer alignment with current legal manned aviation rules and operational oversight requirements. This book has been written with the aim of helping anyone involved in drone operations to develop structures and practices capable of adapting to these changes, based upon the author's experience of over 55 years working in the Aviation Industry and adapting and applying many of the principles of working in a safety critical organisation, to help you apply current best practices to develop your own organisation's Integrated Safety Management System.

This is not meant to be simply an instruction manual but a record of my own personal shared observations, offered in the hope that it will stimulate the reader to ask questions of their own perceptions and values to design, develop, and constantly evolve their own Safety Management System (SMS), particular and appropriate for their specific operations.

We will make many references and suggest links to case studies and training media for the reader to explore and consider at their own convenience. The subject is of such vast proportions that it is impossible to include everything, so I have referenced some topics which in my own humble opinion relate most practically to teaching, understanding, applying, and developing a thirst for more investigative knowledge.

This is not meant as an academic book but a practical guide to help the reader ask appropriate questions enabling a deeper working knowledge of Integrated Safety Management Systems, Human Resource Management, Training, and Development of the knowledge, skill, and attitudes/behaviours to maximise the safety and efficiency of their own organisation.

Throughout my career, I have made many errors and poor decisions but always tried to come to conclusions based upon some very basic principles:

- Is it legal?
- Is it professional?
- Is it morally and ethically right?
- If there is any doubt, there is no doubt!

Preface

Author Introduction

The author has worked over 55 years in the *Aviation Industry* beginning with the Air Training Corps as a young boy and ending a fascinating career, flying as a Senior Airline Training Captain and Flight Simulator Instructor with over 16,500 flying hours flown on 32 different aircraft types ranging from light single-engine trainers to heavy jets such as the Boeing 727 and Airbus A300 B4. I have visited over 300 different airfields, ranging from London Heathrow to extremely short rough airstrips in the African Bush and operated in over 70 countries worldwide.

When I reflect upon my career, I think it turned out not too bad for someone who was told when failing the entry selection for the Royal Air Force, "You obviously do not have officer qualities. Anyway, you chaps from the Newcastle are meant to be building ships or digging coal, aren't you? There is no way you could be a pilot with that Geordie accent!".

So, it shows how much they knew!

Leaving school at 16 with few qualifications after the disappointment of not being suitable for the RAF, I worked at several jobs, often two or three at a time, to finance my desire to fly and at least obtain my Private Pilot's Licence. I worked as a labourer on large construction projects including a spell underground in the sewer systems, barman, selling a range of products from tea, coffee, ice cream, double glazing, solar panels and running a guest house while working as a loader in a fertiliser factory in Chester and completing my Assistant Flying Instructors Rating during the weekends. After several years as a PPL Flying Instructor, I was sponsored by a local Air Taxi Operator flying PA 23 and 31.

Over the following years, I worked extensively in Africa, flying food and medical aid in the wars in Angola and Mozambique with the International Committee of the Red Cross. I later progressed to working with several well-known scheduled airlines and flew diverse operations, from VIP small jet charters to international freight operations to being a personal Pilot to an African president. This later position afforded me the dubious opportunity to spend several nights in the secret desert bunker of the former Libyan leader Muammar Gaddafi and chat at breakfast while staying in King Fahad's Palace in Riyad, Saudi Arabia, with a small white-haired gentleman sandwiched between the two largest bodyguards I have ever seen. I later discovered he was Mahmoud Abbas, head of the Palestine Liberation Organisation!

Now, I am too old and doddery to fly but still with a wealth of knowledge, experience (inside and outside of aviation), and a burning passion for anything that flies, so I have turned to the classroom. I taught Crew Resource Management as CRM Manager for a North Sea Helicopter Operator for six years until redundancy cut short this career change.

Now, I am applying my talents as a freelance CRM trainer/examiner with a new helicopter operator and have joined the exciting and rapidly expanding drone industry, working as a ground instructor with the drone specialist, heliguy™, based in North Shields.

Inspired by this young, dynamic, and innovative company, I have co-written a TQ UK Level 5 training course aimed at Drone Operations Management.

I am very proud and grateful to have written this book in collaboration and with help from Mark Blayney, the Head of Training at heliguy, and colleagues from the Applied Psychology and Human Factors Group, Aberdeen University, Aberdeen, Scotland, UK, where I am working as Industrial Associate.

As part of the many developments in the drone industry, it is predicted that operations will be required to adapt to more stringent regulations very similar to those currently mandated by aviation authorities.

In many cases, there is no reason to reinvent the wheel, and drone operators should take advantage of the vast developments in Safety Management Systems and Human Factors in manned aviation; and this book aims to work as a practical guide to the reader by suggesting and applying some of the tried and tested structures already considered to be best practice.

I hope you enjoy this guide and use it to stimulate your thirst for knowledge and to never stop asking questions to help you improve the frequency and quality of your continuous self-analysis to improve your organisation's and your own personal safety.

Geoff Bain

Co-Author Introduction

HELIGUY™ Head of Training (Accountable Manager) Mark Blaney had 13 (16) years of experience in the drone industry.

- A fixed-wing specialist, Mark deployed the H450 in Afghanistan for the British Army and has worked closely with organisations in Kosovo and Hong Kong.
- Mark's vast experience can be transferred to enterprise drone programmes and has delivered HELIGUY training to the likes of Network Rail and UK emergency services.
- His comprehensive fixed-wing CV makes him ideal to deliver heliguy fixed-wing course.

Upon the completion of his British Army BARB Test, Mark Blaney selected the role of UAS Operator on the hunch that it sounded interesting.

Little did he know that this decision would be the start of a career in the drone industry which would span more than a decade—and counting, taking him from Helmand to heliguy. During this time, the fixed-wing specialist has built an impressive and comprehensive CV, which includes the deployment of the Hermes 450 tactical UAV in Afghanistan for the British Army; working with MAT Kosovo to utilise drones for mine clearance; and clocking in excess of 1,500 (2000) hours of airtime.

Mark's expertise is particularly suited to mentoring Pilots from enterprise drone programmes—a core element of heliguy's training provision. Recent closed-course, industry-specific training that he has delivered to Network Rail, National Grid, and North Wales Police is a case in point.

Mark Blaney

Co-Author Introduction

Target Audience and Overview

- Anyone working in a Safety Critical Industry, needing practical guidance for organising their drone operations
- Any organisation interested in developing a structured approach to their business operations
- The content is based upon best practice used extensively in the manned Aviation Industry
- The content is directly adaptable to *any* industry and can easily be appropriately modified
- The main focus is suggested applications for the rapidly developing UAS Industry
- RPAS Pilots and crew members
- RPAS Accountable Managers and Training Pilots
- Anyone in any role working in this discipline

This book has been written in collaboration and courtesy of the drone specialist, **heliguy**.

It represents several extracts from their **TQ UK Level 5 Drone Operations Managers** course, written and delivered by Mark and myself. This course highlights the most advanced aspects of subject matter knowledge required to plan, execute, deliver, and monitor a professional drone operation. It is in no way meant as a substitute for their Operations Managers course mentioned above, which involves 5 days of classroom training, followed by a 12-week timeline to review and complete the assignments prior to assessment and award. Successful completion qualifies as an equivalent to a foundation degree. This guide has been organised in a specific sequence, introducing the reader to a logically developing progression of the following:

Part One—Integrated Safety Management Systems: Highlights aviation processes which can be applied to any business operation. These are the basic concepts of a Safety Management Systems structure as developed for manned aviation.

Part Two—Human Factors, Crew Resource Management, Team Resource Management: This part is separated into two parts, **Generic** and **Job Specific**.

The former represents a suggested syllabus that is equally applicable to anyone in any industry to stimulate an interest in improving company safety and harmony by encouraging an interest in understanding individual and group behaviour and the effects upon safety and productivity (**referred to as TRM**).

The latter represents the additional job-specific subject matter which should be understood by all employees in their respective sector of employment. The example highlighted is related to those employed as Pilots and engineers in the Aviation Industry as a mandatory requirement known as CRM.

Note: In commercial aviation, Pilots and engineers would be mandated to study both sections to maintain legal compliance and competence.

Part Three—Accident/Incident Investigation/Case Studies: In this part, we will investigate several case studies of various accidents and incidents where the reader can apply any new skills to analyse and then suggest root causes and suggested mitigations (corrective actions), comparing their own opinion to what occurred. We have deliberately selected five case studies from different industries since the analytical process is identical. What is probably most interesting is the fact that even though the actual risks may be different, the outcomes, corrective actions and conclusions hardly vary between such diverse operations.

Part Four—Education: One of the fundamental philosophies of this book is the importance of the fact that all employees working in Safety Critical operations receive regular, structured, and appropriate professional training to help them understand, develop, and maintain their required competencies in the workplace. Therefore, it is a logical requirement that everyone in authority should understand the basic mechanics of teaching and learning. Even if they are not directly employed as a trainer, at least management and all employees should understand the knowledge required to perform their role.

Part Five—Leadership and Management: This part is designed to deconstruct some of the basic competencies of Leadership and Management. It is based upon the philosophy that all organisations have leaders and managers with specific accountabilities and responsibilities. To perform their tasks safely and cost effectively, they need to understand some basic principles of a company's culture and individual understanding and adoption of that culture as well as understanding principles of human behaviour and how their own behaviour can impact upon the group behaviour. What is often overlooked is the role of the **Follower** in this set up. They should have the same access to knowledge and understanding as the **Leader** with the chance to aspire to leadership themselves.

The only real difference between a Leader and Follower is that the former has the casting vote in final decision-making.

Part Six—Practical Application in the RPAS (Drone) Industry: In this part, we will look at some of the basic practical processes to help you organise your own SMS (Drone Operations Manual). We will discuss the importance and process of regulatory as well as internal and external auditing to maintain quality assurance.

Part Seven—Research: This small part highlights some of the research being carried out to suggest improvements in organising and operating within an exciting and rapidly exponentially expanding industry.

Part Eight—Resources, Links, and Recommendations: This part contains additional material to complement the book and allow the reader to explore this vast subject in greater depth if they wish to obtain a much greater depth of knowledge in areas including:

- Regulatory authorities
- Suggested reading material sources

- Reference manuals
- Bibliography
- Online hyperlinks
- Video media
- Relevant organisations

Acknowledgements

I would like to acknowledge and thank the following for their help, support, and inspiration:

HELIGUY

- Mark Blaney—Accountable Manager and former Head of Training
- James Willoughby—Blogger, Drone Content Executive
- Ruairi Hardman—Business Development Manager
- Jack Sharp—Head of Training -Photographer

APPLIED PSYCHOLOGY AND HUMAN FACTORS GROUP, ABERDEEN UNIVERSITY

- Professor Rhona Flin,
  Professor of Industrial Psychology at Aberdeen Business School
  Robert Gordon University, Scotland, UK
- Dr Amy Irwin
  Senior Lecturer and Lead of Applied Psychology and Human Factors Group, University of Aberdeen, Aberdeen, Scotland, UK
- Dr Oliver Hamlet
  Axiom Human Factors Ltd, Applied Psychology and Human Factors Group
  University of Aberdeen

UNIVERSITY COLLEGE LONDON

- Professor Robert J. West
  Professor of Health Psychology
  Health Behaviour Research Centre
  Department of Epidemiology and Public Health

SKYDIO

- Ben Shirley—EMEA Solutions Engineer

UK CIVIL AVIATION AUTHORITY

- Callum Conde (previously Holland) RPAS Sector Lead

Development

DRONE INDUSTRY TO GROW TO "£29.8 BN BY 2026"

PUBLISHED ON 13 AUGUST 2021

The drone market will grow by 9.4% over the next five years to become a £29.8 billion industry by 2026, according to a new report. The study shows the energy industry has the highest adoption rates, while mapping and surveying will remain the top application of drones.

These are the findings in the recently published Drone Market Report 2021–2026, compiled by German company Drone Industry Insights (Droneii).

Figure 0.2, taken from the findings, shows the current size of the market, and how it is projected to rise.

The report also shows that commercial drones are being used in a vast majority of industries, and the use of enterprise UAS will continue to climb. Other key takeaways from the report are: Drone applications in the energy industry are on path to earn just under £4.26 billion throughout the globe.

Other industries such as construction and agriculture are not far behind, and some industries related to warehousing and insurance will grow at a more rapid pace in the next five years. Drone services, such as mapping and inspections among dozens of others, represent roughly 78% of global drone-related revenue and are the main

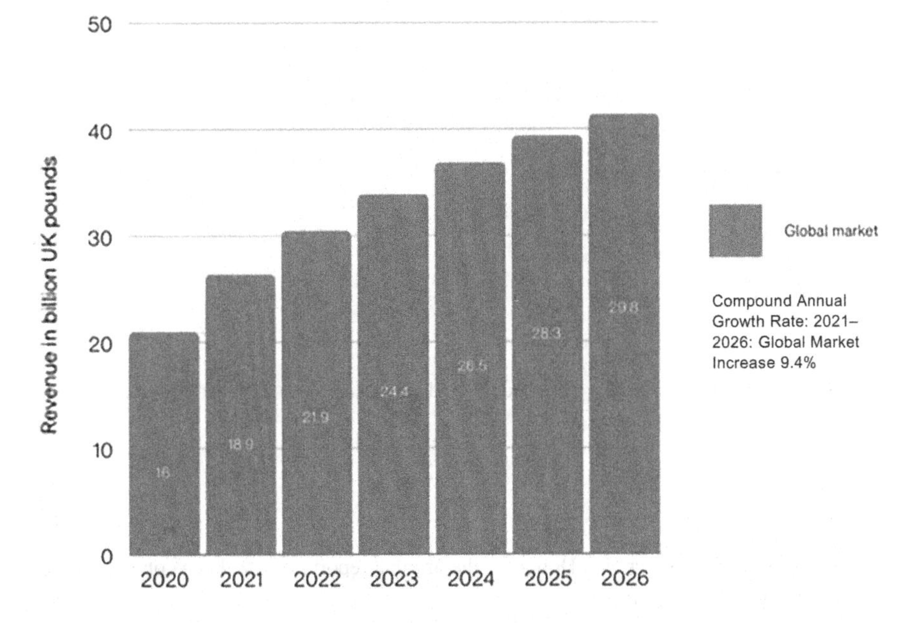

FIGURE 0.2 Revenue from drones in £ millions against the average projected expansion rate of 9.4% per annum between 2020 and 2026.[1]

driving force in the market. Some of these also experienced a positive impact of the pandemic by assisting with the remote delivery of coronavirus test kits and vaccines that allowed people to keep a safe distance and avoid infection.

Drone hardware is forecasted to experience strong growth. Unit sales will grow from 828,000 in 2021 to almost 1.4 million in 2026 at 10.6% CAGR (Compound Annual Growth Rate). The commercial drone market today is led regionally by Asia thanks to China and Japan, while South America and India are growing the fastest at the regional and country levels, respectively. The report, based on proprietary data and market models supported by primary and secondary sources, is the latest edition of Droneii's yearly study on the global drone market. It analyses the drone industry by segment (hardware, software, services), by industry sector (energy, construction, agriculture, etc.) and by application method (e.g. mapping and surveying, inspection, and delivery, among others) and provides a more in-depth look into all regional markets as well as a deeper analysis of the top ten countries for drones.

PROJECTED NUMBER OF COMMERCIAL DRONES BY INDUSTRIAL SECTOR IN 2030

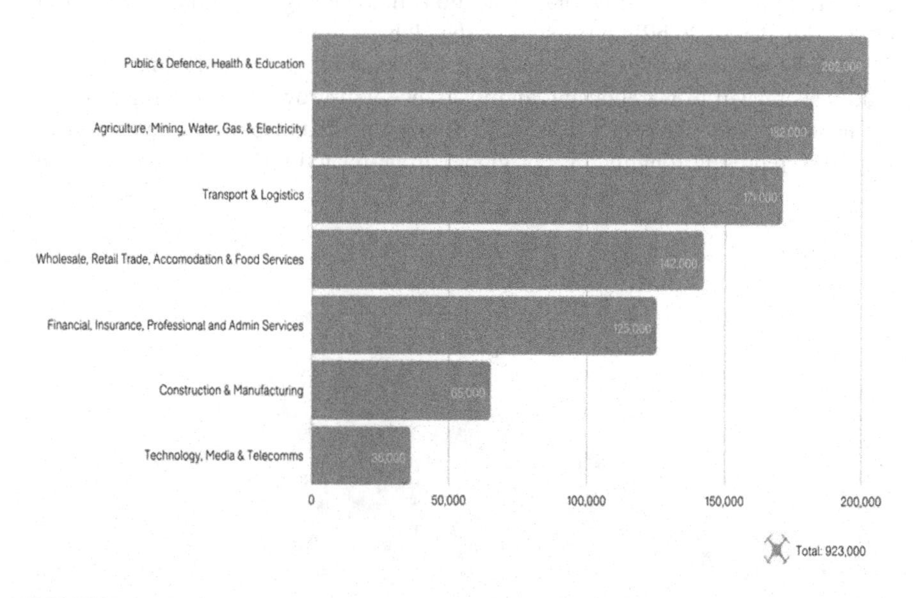

FIGURE 0.3 Projected number of drones by industrial sector in 2030 estimated as a total of 923,000.

NOTE

1 The monetary figures were US Dollars in the original report but have been converted to UK pound sterling for this book.

Acronyms and Abbreviations

AAIB	Air Accidents Investigation Branch
ADS-B	Automatic Dependent Surveillance Broadcast
AGL	Above Ground Level
AIAA	Area of Intense Air Activity
AIP	Aeronautical Information Publication
AIS	Aeronautical Information Service
ALARP	As Low As Reasonably Practicable
AltMOC	Alternate Means of Compliance
AMC	Acceptable Means of Compliance
AMSL	Above Mean Sea Level
ANO	Air Navigation Order
APU	Auxiliary Power Unit
ATC	Air Traffic Control
ATCU	Air Traffic Control Unit
ATZ	Aerodrome Traffic Zone
AUP	Airspace User Portal
BMFA	British Model Flying Association
BVLOS	Beyond Visual Line of Sight
CAA	Civil Aviation Authority
CAP	Civil Aviation Publication
Con-Ops	Concept of Operations
CPL	Commercial Pilot Licence
CRM	Crew Resource Management
CTA	Controlled Traffic Area
CTR	Controlled Traffic Region
CU	Command Unit
C2	Command and Control
DA	Danger Area
DAA	Detect and Avoid
EASA	European Aviation Safety Agency
ECCAIRS	European Co-ordination Centre for Accident and Incident Reporting System
ENSF	Enhanced Non-Standard Flight
ERF	Emergency Restriction of Flying
ESC	Electronic Speed Controller
EU	European Union
EVLOS	Extended Visual Line of Sight
FIR	Flight Information Region
FISO	Flight Information Service Officer
FL	Flight Level
FOD	Foreign Object Damage/Debris
FPV	First Person View

FRTOL	Flight Radio Telephony Operators Licence
FRZ	Flight Restriction Zone
GNSS	Global Navigation Satellite System
GPS	Global Positioning System
GUI	Graphical User Interface
HIRTA	High-Intensity Radio Transmission Area
HMI	Human-Machine Interface
ICAO	International Civil Aviation Organisation
IOSD	Intelligent On-Screen Display
IMU	Inertial Measurement Unit
JARUS	Joint Authorities for Rulemaking on Unmanned Systems
Li-Po	Lithium Polymer
LS	Landing Site
MAA	Military Aviation Authority
MATZ	Military Aerodrome Traffic Zone
METAR	Meteorological Terminal Aviation Routine
MOR	Mandatory Occurrence Reporting
MTOM	Maximum Take-off Mass
NAA	National Aviation Authority
NATS	National Air Traffic Services
NOTAM	Notice to Airmen
NSF	Non-Standard Flight
OM	Operations Manual
OS	Ordnance Survey
PDRA	Pre-Defined Risk Assessment
PfCO	Permission for Commercial Operation
PIB	Pre-Flight Information Bulletin
PMU	Power Management Unit
RA	Risk Assessment
RAE	Recognised Assessment Entity
RAMS	Risk Assessment Method Statement
RA (T)	Restricted Area (Temporary)
RP	Remote Pilot
RPAS	Remotely Piloted Aircraft System
RPZ	Runway Protection Zone
RTH	Return to Home
SI	Statutory Instrument
SOP	Standard Operating Procedure
SORA	Specific Operations Risk Assessment
SPOF	Single Point of Failure
STS	Standard Scenario
TAF	Terminal Aerodrome Forecast
TDA	Temporary Danger Area
TOLS	Take-Off and Landing Site
TRM	Team Resource Management

UA	Unmanned Aircraft
UAS	Unmanned Aircraft System
UAS OSC	Unmanned Aircraft System Operating Safety Case
UKAB	United Kingdom Airprox Board
VFR	Visual Flight Rules
VLOS	Visual Line of Sight

Definitions

Accident—An occurrence associated with the operation of an aircraft which, in the case of an unmanned aircraft, takes place between the time the aircraft is ready to move with the purpose of flight until such time when it comes to rest at the end of the flight and the primary propulsion system is shut down, in which:

a) a person is fatally or seriously injured as a result of being in the aircraft or having direct contact with any part of the aircraft, including parts which have become detached from the aircraft or having direct exposure to jet blast, except when the injuries are from natural causes, self-inflicted or inflicted by other persons, or when the injuries are to stowaways hiding outside the areas normally available to the passengers and crew; or

b) the aircraft sustains damage or structural failure which adversely affects the structural strength, performance or flight characteristics of the aircraft and would normally require major repair or replacement of the affected component, except for engine failure or damage, when the damage is limited to a single engine (including its cowlings or accessories) to propellers, wing tips, antennas, probes, vanes, tyres, brakes, wheels, fairings, panels, landing gear doors, windscreens, the aircraft skin (such as small dents or puncture holes) or minor damages to main rotor blades, tail rotor blades, landing gear and those resulting from hail or bird strike (including holes in the radome); or

c) the aircraft is missing or is completely inaccessible.

Accountable Manager—A nominated person who has the authority for ensuring that all activities are carried out in accordance with the applicable requirements and regulations. The Accountable Manager is also responsible for establishing and maintaining an effective Management System.

Aircraft—Any machine that can derive support in the atmosphere from the reactions of the air.

Airprox—An airprox is a situation in which, in the opinion of a Pilot or air traffic services personnel, the distances between aircraft as well as their relative positions and speed have been such that the safety of the aircraft involved may have been compromised.

Assemblies of People—Gatherings where persons are unable to move away due to the density of the people present.

Autonomous Operation—An operation during which an unmanned aircraft is operating without Pilot intervention in the management of flight.

Beyond Visual Line of Sight—A type of UAS operation which is not conducted in VLOS.

Certified Category—Operations that present an equivalent risk to that of manned aviation and so will be subjected to the same regulatory regime (i.e. certification of the aircraft, certification of the operator, licencing of the Pilot).

Command and Control Link—The data link between the remotely piloted aircraft and the Command Unit to manage the flight.

Concept of Operations—Describes the characteristics of the organisation, system, operations and the objectives of the user.

Danger Area—Airspace which has been notified as such within which activities dangerous to the flight of aircraft may exist at specified times.

Detect and Avoid—The capability to see, sense or detect conflicting traffic or other hazards and take the appropriate action.

Extended Visual Line-Of-Sight Operation—An operation in which the RP and RPA Observer(s) maintain direct unaided visual contact with the RPA sufficient to monitor the aircraft flight path in relation to other aircraft, persons, vessels, vehicles and structures to avoid collisions.

Graphic User Interface—A method of inputting data and commands into a computer by interacting with graphical icons on a screen, either by touch or via a peripheral device. It may also output data to the user.

Handover—The act of passing piloting control from one RP to another.

Highly Automated—Systems that still require inputs from a human operator but which can implement the action without further human interaction once the initial input has been provided.

Intelligent On-Screen Display—The Graphic User Interface element for a DJI-manufactured UA.

Involved Persons—A person may be considered to be involved when they have given explicit consent to the UAS operator or to the Remote Pilot to be part of the UAS operation (even indirectly as a spectator or just accepting to be overflown by the UAS) and received from the UAS operator or from the Remote Pilot clear instructions and safety precautions to follow in case the UAS exhibits any unplanned behaviour. In principle, in order to be considered a "person involved", one must able to decide whether or not to participate in the UAS operation; broadly understands the risks involved; has reasonable safeguards during the UAS operations, introduced by the site manager and the aircraft operator and is not restricted from taking part in the event or activity if they decide not to participate in the UAS operation.

Knot—A unit of speed equal to one nautical mile (1.852 km) per hour, approximately 1.151 mph.

LOSA—Line Operational Safety Audit

Lost Link—The loss of command-and-control link contact with the remotely piloted aircraft such that the Remote Pilot can no longer manage the aircraft flight.

Maximum Take-Off Mass (MTOM)—MTOM is the maximum mass defined by the manufacturer or the builder, in the case of privately built UAS, which ensures the controllability and mechanical resistance of the UA when flying within the operational limits. The MTOM should include all the elements on board the UA: (a) all the structural elements of the UA; (b) the motors; (c) the propellers, if installed; (d) all the electronic equipment and antennas; (e) the batteries and the maximum capacity of fuel, oil and all fluids and (f) the heaviest payload allowed by the manufacturer, including sensors and their ancillary equipment.

Nautical Mile—A unit of distance that is approximately one minute of arc of latitude (1.15 miles = 1 nautical mile).

Night—The hours between the end of evening civil twilight and the beginning of morning civil twilight.

Open Category—Operations that present a low (or no) risk to third parties. Operations are conducted in accordance with basic and pre-defined characteristics and are not subject to any further authorisation requirements.

Operational Authorisation—A document issued by the CAA that authorises the operation of an unmanned aircraft system, subject to the conditions outlined within the authorisation, having taken into account the operational risks involved.

Pre-Flight Inspection—The inspection carried out before the flight to ensure that the aircraft is fit for the intended flight.

Remote Pilot—A natural person responsible for safely conducting the flight of an unmanned aircraft by operating its flight controls, either manually or, when the unmanned aircraft flies automatically, by monitoring its course and remaining able to intervene and change the course at any time.

Serious Incident—An incident involving circumstances indicating that there was a high probability of an accident and associated with the operation of an aircraft which, in the case of an unmanned aircraft, takes place between the time the aircraft is ready to move with the purpose of flight until such time when it comes to rest at the end of the flight and the primary propulsion system is shut down.

Safety—The state in which risks associated with aviation activities, related to or in direct support of the operation of aircraft, are reduced and controlled to an acceptable level.

Safety Management System—A systematic approach to managing safety, including the necessary organisational structures, accountabilities, policies and procedures.

Specific Category—A category of UAS operations that is described in Article 3 of the UAS Implementing Regulation.

UAS Delegated Regulation—Commission Delegated Regulation (EU) 2019/945 on unmanned aircraft systems and on third-country operators of unmanned aircraft systems, as retained in UK domestic law.

UAS Implementing Regulation—Commission Implementing Regulation (EU) 2019/947 on the rules and procedures for the operation of unmanned aircraft, as retained in UK domestic law.

UAS Operator—Any legal or natural person operating or intending to operate one or more UAS.

Uninvolved Person—Persons who are not participating in the UAS operation or who are not aware of the instructions and safety precautions given by the UAS operator.

Unmanned Aircraft—Any aircraft operating or designed to operate autonomously or to be piloted remotely without a Pilot on board.

Unmanned Aircraft Observer—A person, positioned alongside the Remote Pilot, who, by unaided visual observation of the unmanned aircraft, assists the Remote Pilot in keeping the unmanned aircraft in VLOS and in safely conducting the flight.

Unmanned Aircraft System—An unmanned aircraft and the equipment to control it remotely.

Visual Flight Rules—Procedures to navigate and avoid collisions without navigational instruments.

Visual Line-of-Sight Operation—An operation in which the Remote Pilot or UA observer maintains direct unaided visual contact with the aircraft.

Part One

Integrated Safety Management Systems

FIGURE PI.1 DJI Matrice 300 flying near woodland on a dark cloudy day with heavy rainclouds nearby.

DOI: 10.1201/9781032620220-1

1 Safety Management Systems

1.1 LEARNING OUTCOMES

By the end of the lesson, students will be able to:

LO 1: Understand the basic principles of Safety Management Systems.
LO 2: Describe how Safety Management principles can be applied to RPAS Operation.
LO 3: Understand the importance of hazard management for RPAS Operations.
LO 4: Analyse the four main components of Safety Management Systems.

1.2 WHAT IS A SAFETY MANAGEMENT SYSTEM (SMS)?

International Civil Aviation Organisation (ICAO) definition/reference ANNEX 6 ICAO Doc 9859: ***A systematic approach to managing safety, including the necessary organisational structures, accountabilities, policies and procedures.***

All Safety Critical Operations by their nature are conducted in environments where the potential for catastrophic and possible life-threatening danger is always present. It is not possible to prevent all these events completely, but it is possible to greatly reduce their probability by approaching all operations with standardised, structured and organised methodologies. Integrated Safety Management Systems (ISMS or SMS) are a tried and tested structure, scientifically designed with over 50 years' evolution and pioneered by operations such as in the Aviation Industry.

Nowadays, in manned aviation, SMS is a legal requirement. In the European Union Aviation Safety Agency (EASA) regulation ORO GEN. 200 , this mandates that operators must have their own SMS based upon the compliance with these legal rulings. The structure highlighted in this book is legally compliant with current Aviation Industry standards and regulations and has been used as the example template since it represents the most advanced and innovative pioneering sector within many of the Safety Critical Industries. Although based upon the Aviation Industry, it can be easily adapted as appropriate by any other operation and specifically with so many principles being directly transferrable to the RPAS Industry. Wherever there is a human dimension or involvement in a process, there will be complex potential integration issues with the MAN/MACHINE interface. This publication suggests tried and tested methods of mitigation. Although most of these recommendations are generic and applicable to all Safety Critical Operations, they are based upon current

DOI: 10.1201/9781032620220-2

FIGURE 1.1 US Federal Aviation Administration diagram showing the relationship between psychological, behavioural, and system/environmental factors in a Safety Management System (compiled by US Federal Aviation Administration (FAA) entitled Safety Management System Fundamentals).

Aviation Industry best practice and specifically relevant to the rapidly developing RPAS Industry. Many believe that as RPAS operations develop and expand at an exponential rate, it will be logical for them to become more regulated and align with many of the current practices mandated in manned aviation. Safety Management Systems rely heavily upon clearly defined structural practices but particularly upon the management team and all the workforce having an in-depth understanding of Human Factors (HFs) or Crew Resource Management (CRM) and how to recognise and understand the relationships between individual and group human behaviour and how that impacts upon safety and efficiency. To put these factors into more context, consider the relationships between the various aspects of an idealised aviation business structure as represented in the slide given in Figure 1.1. The cycle represents a continuum where you cannot influence one aspect in isolation without there being an impact upon the other factors.

1.3 WHY DO I NEED A SAFETY MANAGEMENT SYSTEM?

- It is a legal requirement.
- It improves the understanding of the causes of accidents and incidents.
- It structures and organises the processes of how to recognise, classify the severity, document and mitigate occurrences.
- Train organisation members.

Operating within a structured Safety Management System provides several advantages. Apart from being a legal compliance requirement, it creates an improved understanding as to why accidents and incidents are caused. Without such a system, an organisation's ability to perform such monitoring would be random and driven more by good luck than good management. An SMS structures and organises processes to recognise and classify severity of occurrences and document their details.

With this information and the establishment of an appropriate database, it is much easier to understand the cause and mitigate occurrences.

Investigation of this data facilitates the development of suitable training programmes for organisation members.

1.4 WHAT FACTORS DICTATE THE STRUCTURE OF A SAFETY MANAGEMENT SYSTEM?

It is an organised approach to managing safety by identifying a standardised, organisational structure. An SMS clearly defines the accountabilities and responsibilities of key personnel by having a transparent "Bill of Rights", identifying the relationships and interaction between management and workforce.

Effectively, it is a contract between the management and employees where everyone should be able to understand what is expected of each organisation member in terms of standards, behaviour and culture, defining what everyone can expect from the organisation and what the organisation can expect from its employees.

Practically, an SMS streamlines the identification, assessment and measurement of threats, errors and predicted hazards so that mitigating strategies can be developed and implemented.

A robust SMS aims to reduce risks to a level **As Low As Reasonably Practicable (ALARP)**.

ALARP specifies an acceptable level of risk rather than complete elimination.

This is an interesting concept which is founded on the following:

The only way to avoid aviation accidents and incidents completely is not to fly at all!

Therefore, operating in any environment will carry associated inherent risks.

If the decision is then to conduct operations, statistically, there will be accidents and incidents of varying severity at some point.

The commercial reality (often determined by the insurance industry) is what level of mitigating investment will be required to decide how safe an operation can be expected to be.

1.5 WHAT FACTORS DETERMINE THE COMPLEXITY OF AN ORGANISATION?

- Number of employees
- Number and complexity of the equipment operated
- Number of bases
- Number of aircraft movements

- Number of approvals or exemptions held by the organisation
- Environmental factors

Your organisation may already perform and comply with many of the elements of an ISMS (Integrated Safety Management System) but may need to organise them in a more structured way. There is no standard "one-size-that-fits-all" SMS.

The priority is to develop an SMS which is effective and works. Key components of an SMS are described in the next section.

1.6 THE FOUR SAFETY MANAGEMENT SYSTEM FUNCTIONAL COMPONENTS

1.6.1 Safety Policy and Objectives

- Management commitment and responsibility
- Safety accountabilities
- Appointment of key staff members
- Emergency Response Planning
- SMS documentation

1.6.2 Safety Risk Management

This is a process using standardised structured methods to identify hazards, risk assess and mitigate them; that is

- Hazard identification
- Risk Assessment and mitigation

1.6.3 Safety Assurance

This process is the method of performance monitoring and measurement. Management of Change is a structured approach to how an organisation deals with and learns how changes can be managed and how they will impact upon the company performance. In any dynamic industry, change is a continuous process, and it must be managed methodically and not left to random evolution. It is a method of using past and present experiences to predict future trends.

1.6.4 Safety Promotion

All company employees need to be kept in the loop as to how safety and efficiency can be achieved, maintained and improved. Promoting a strong safety culture is a fundamental tool in this process. The following are fundamental for achieving these:

- Training and education
- Safety communication

2 Safety Policy and Objectives

2.1 MANAGEMENT COMMITMENT AND RESPONSIBILITY

Senior management should use the safety policy to make a clear commitment to ensure that they allocate sufficient resources, time, and responsibilities. All the previously discussed requirements of an SMS should be committed to through a clearly written policy statement that should be signed by the Accountable Manager. The Accountable Manager should also actively demonstrate commitment to the policy and ensure that all principles are taught by allocating suitable training, understood, and practised by all company members. The Accountable Manager should ensure that a "Just Culture" is encouraged within the organisation. A Just Culture is key to the successful implementation of safety regulations. It should encourage an environment of reporting and will depend on how the organisation handles these reports regarding blame and punishment. The Just Culture should instil an atmosphere of trust in which employees are encouraged to report any safety-related information. However, a clear line must be drawn between what is deemed acceptable and what is unacceptable behaviour.

2.1.1 Principles to Be Included

The following principles should be included in the Accountable Managers Statement:

- Fundamental approach to safety
- Senior management's commitment to safety
- Commitment to provide adequate resources to manage safety and reduce risks to an acceptable level
- Encourage all organisation members to actively participate in and fulfil all aspects of SMS
- Promote, encourage, and monitor organisational Just Culture

There is an excellent Just Culture video produced by the UK CAA available on YouTube:

See Part Eight.

The Most Important Concept of This Publication

Just Culture is probably the most important concept in this publication, and therefore I have inserted a great deal of background information and references for further research contained in Part Eight.

DOI: 10.1201/9781032620220-3

Flight Safety Organisation
A Roadmap to a Just Culture: Enhancing the Safety Environment **is a very well recommended read.**

2.2 SAFETY ACCOUNTABILITIES

2.2.1 Management Structure

It is essential that the management structure should be clearly defined so that all company employees and external entities such as customers and official regulators have a clear point of reference/contact with a transparent understanding of who is responsible and accountable for the associated business functions and an understanding of reporting lines and hierarchy. The usual form of presentation is with an organogram. The Accountable Manager has the ultimate legal and administrative power to monitor and ensure the Company Standards are maintained. Their duties may be delegated to other suitable organisation personnel, but the Accountable Manager always assumes the ultimate accountability for safety. Having a clear written statement of Safety Policies and Objectives facilitates a successful SMS.

2.2.2 Key Staff Members (Nominated Persons)

Appropriate nominated persons (NPs) may be appointed with specific roles to support the SMS by supervising those appropriate roles as delegated by the Accountable Manager. These functions are often performed by a Safety Committee comprising appropriate subject matter experts who are capable of interfacing with external groups or organisations in addition to their own organisational colleagues. They should all meet at regular intervals to discuss the monitoring review update and continuous development of safety-related issues. In an efficient operation, particularly with reference to safety critical aspects, it is essential for everyone to know and understand their respective duties and who the subject matter experts are with specific qualifications and abilities. This is in addition to knowing what are the correct lines of communication.

These roles are considered as Key Roles and are often classified as to be for nominated persons. Given next is a typical organogram.

These are the minimum roles that would be nominated in a standard UAS organisation. The Accountable Manager and Remote Pilot as a minimum, and UA Observer and Camera Operator would be optional.

If additional roles were added, such as a Flight Safety Manager or Chief Pilot, then these roles should also be nominated. Nota bene (NB): In many organisations, the Flight Safety Manager is required to have a direct line of access to the Accountable Manager.

2.2.3 Emergency Response Plan (ERP)

In a safety critical industry, it is no good just reacting in a random fashion when accidents/incidents occur.

Using past and present data to help predict future potential hazards, a much higher statistical probability of taking correct mitigating actions can be developed.

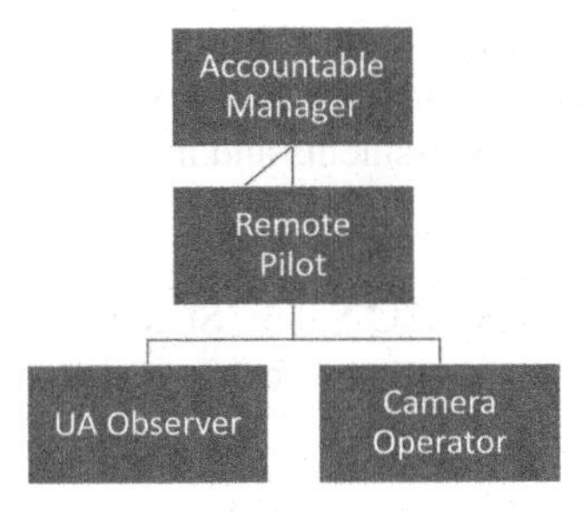

FIGURE 2.1 An organigram depicting a typical organisation structure of a small operator.

Continuous monitoring review, update, and re-training are the founding principles of developing a good ERP reaction process.

Emergency Response Plans need to be clearly understood by all organisation members with a particular attention to individual and group roles. Who does what, how, and when?

A plan designed to describe staff actions in an emergency should include the following:

- Orderly transition from normal to emergency operations
- Designated emergency person in authority
- Designation and assignment of emergency roles, responsibilities, and accountabilities
- Coordination structures to resolve emergencies
- Safe continuation of operations or return to normal operations as soon as possible
- Identification of staff roles and actions together with the details of any interaction from emergency services or external third-party agencies

ERPs must be regularly practised and reviewed by all participants to maintain competence. They are often a regulatory requirement and should regularly be updated. They could consist of cards or checklists that hold appropriate contact information for nominated persons. They should contain information on specific duties and be made available for nominated persons who are participating in accident- and incident-training exercises.

2.2.4 SMS Documentation and Control

It is essential that appropriate records are maintained with a system of revision and distribution control. Not only will this streamline administrative process, but it also improves any internal or external audit process.

Documentation should include:

- The safety policy and objectives of the SMS
- The accountabilities and responsibilities of the Accountable Manager and key staff members

- Any safety-related processes, procedures, or checklists
- The results and subsequent actions from any safety audits or assessments
- The results of any Risk Assessments and mitigation measures in place
- A hazard log

Below shows us a typical UAS OSC Safety Statement signed by the Accountable Manager. Although the wording will vary between different organisations.

SAFETY STATEMENT

Small Unmanned Aircraft Ltd is committed to safety as the primary consideration while conducting Unmanned Aircraft System (UAS) operations. The protection of employees, contractors, and the general public is of paramount importance, and, as such, all operations will be conducted in a safe and responsible manner.

Small Unmanned Aircraft Ltd embraces "Just Culture" and fosters an open honest environment where employees are encouraged to voice any concerns over safety-related issues so they can be addressed immediately.

The Accountable Manager must ensure that any operation of any UAS is carried out in accordance with and abide by the requirements of ANO 2016–2020 Amendment and UAS Implementing Regulation 2019/947 (as retained in UK Law), the conditions of this Operating Manual, the relevant PDRA, insurance policy, and the Operations Authorisation issued by the Civil Aviation Authority (CAA).

To ensure the safest operations are maintained, Small Unmanned Aircraft Ltd shall implement a safety ethos by:

- Ensuring a safe working environment for all employees, contractors, and the public
- Operating UAS in a safe and responsible manner in accordance with the legislation and the conditions detailed in the Operational Authorisation
- Ensuring that all personnel are suitably trained, current, and competent to participate in operations
- Planning, assessing, and executing all operations in accordance with the procedures outlined in this Operations Manual
- Managing risk and implementing suitable mitigation where required
- Ensuring that the personnel do not deviate from the procedures outlined within this Operations Manual unless acting in an emergency where the relevant emergency procedures must be followed
- Reporting any accident, serious incident, reportable occurrence, or Airprox in a timely manner to the appropriate authority
- Conducting equipment maintenance in accordance with the manufacturer's recommendations
- Keeping up to date with changes in legislation by way of the CAA website and Skywise platform, ensuring this document remains compliant at all times

FIGURE 2.2 Sample mission statement of an organisation needed as part of the Safety Management System requirements.

Small Unmanned Aircraft Ltd, as the UAS operator hereby confirms that the intended operation(s) as described herein will comply with any applicable rules relating to it, in particular, with regard to privacy, data protection, liability, insurance, security, and environmental protection.

Signed: Joe Bloggs
Accountable Manager
Small Unmanned Aircraft Ltd

FIGURE 2.2 (Continued)

Next, we have an example of a Safety Policy document. This basic example is taken from the UK Government website and from the Health and Safety Executive **(HSE).**

Part 1: Statement of intent

This is the Health and Safety Policy statement of:	
Our Health and Safety Policy is to:	
Signed	**Date**
Print Name	**Review Date**

FIGURE 2.3 Blank template Safety Policy Statement as suggested by the UK Health and Safety Executive (HSE).

Part 2: Responsibilities for Health and Safety

Overall and final responsibility for Health and Safety
Day to day responsibility for ensuring this policy is put into practice
To ensure that Health and Safety standards are maintained/improved, the following people have responsibility in the following areas:
All employees should
• **Cooperate with supervisors and managers on Health and Safety Policies.** • **Take care of their own health and safety; and** • **Report all health and safety concerns to an appropriate person as detailed above.**

FIGURE 2.4 Continuation of Figure 2.3.

Part 3: Arrangements for Health and Safety

Risk Assessment
Training
Consultation
Evacuation

FIGURE 2.5 Safety Policy statement.

3 Safety Risk Management

3.1 OVERVIEW

Having a standardised structured process allows an organisation to classify and make an assessment as to what impact the hazard may have in terms of:

- Safety
- Financial Impact
- Commercial Reputation

Once a hazard has been identified, you can apply the ALARP (As Low As Reasonably Practicable) concept.

ALARP is the principle that not all risks can be eliminated, or it is economically impractical to do so, and all mitigations must be managed in order to bring the risk to a level known as "As Low As Reasonably Practicable (ALARP)".

A decision can then be made as to how severe and tolerable the hazard is and what impact upon the organisation would occur if the hazard re-occurred.

Corrective or remedial mitigating actions can now be implemented, and the consequences of these actions monitored and documented.

The safety risk management process begins with identifying the hazards impacting the safety of your organisation:

- Assess the risks in terms of levels of likelihood and severity
- Once level assessed:
  - Take appropriate remedial action or mitigating measures in line with ALARP.
  - Monitor mitigating actions to ensure performance.

The diagram given in **Figure 3.1** is called the **Safety Cycle**. And it is the foundation of all SMS theory. For this to be effective, it must be a continuous process.

3.1.1 Reporting Systems

Hazards can only be controlled if they are known about, and the implementation of a confidential safety Reporting System facilitates this process. Safety reporting may be:

- Proactive (before the event)
- Reactive (after the event)
- Predictive (trying to predict what may happen)

DOI: 10.1201/9781032620220-4

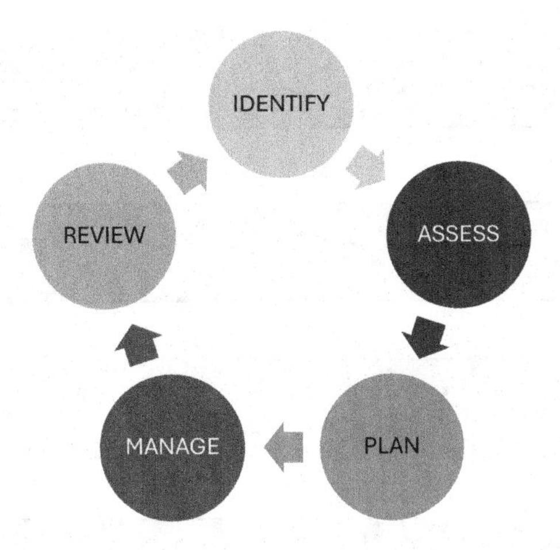

FIGURE 3.1 The Safety Cycle.

3.1.2 Mandatory Occurrence Reporting Scheme

All organisational members should be trained and encouraged to report occurrences and understand:

- What to report?
- How to make a report?
- Who to report it to?

All such data will contribute to monitoring safety performance and trends in order to develop appropriate corrective actions.

3.1.3 Voluntary Occurrence Reporting Scheme

The Voluntary Occurrence Reporting Scheme (VORS) adopts the same format as the Mandatory Occurrence Reporting Scheme (MORS) but encompasses occurrences which remain relevant to aviation safety and fall outside the classification of a MOR.

3.1.4 Confidential Reporting Programme for Aviation and Maritime (CHIRP)

CHIRP's UAS programme was established to provide a confidential safety-related reporting function, independent of MORS and VORS, acknowledging there may be occasions when the reporter has an anxiety of reporting to their employer or the aviation regulator directly. Similarly, individuals who operate in the Open Category, irrespective of the requirement to maintain an SMS and Operations Manual, may choose to utilise the CHIRP reporting programme.

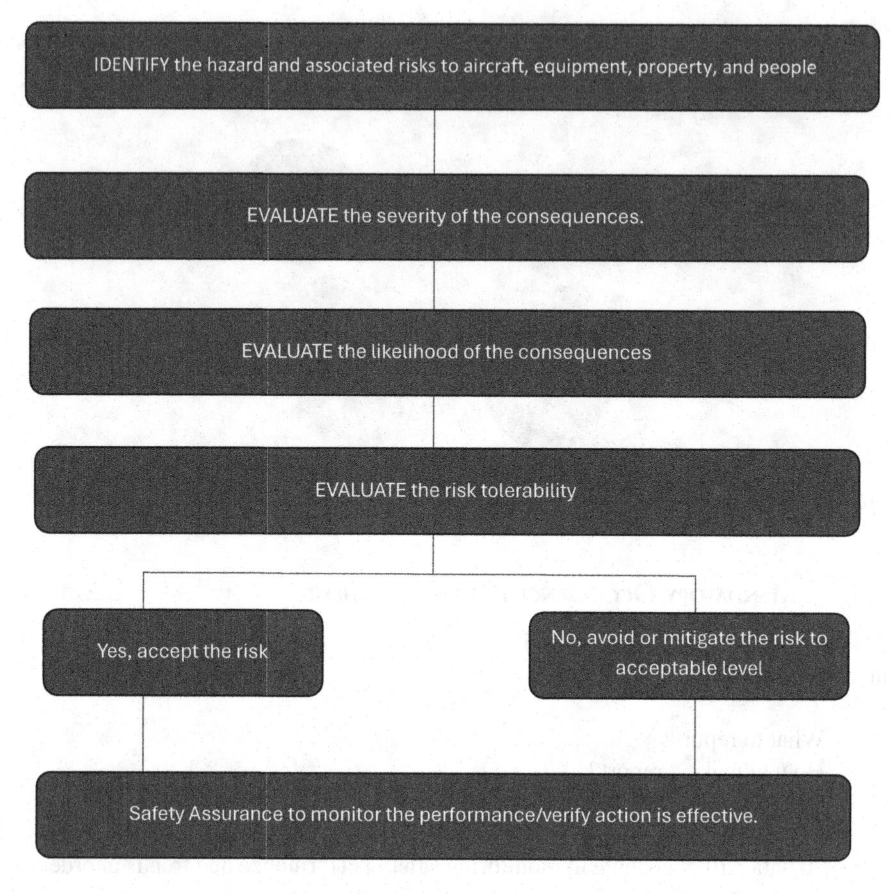

FIGURE 3.2 Hazard Evaluation flow chart.

3.1.5 Risk Assessment and Mitigation

This process facilitates the identification of risk level with associated hazards in terms of potential harm. Such a process is described in the above diagram in Figure 3.2.

3.1.6 Definitions

A **hazard** is defined as: "Something that has the potential to cause harm".

A **risk** is defined as: "The circumstance of someone coming into contact with the hazard together with the consequence of doing so".

Generally, a hazard exists in the present, whereas the risk associated with it is a potential outcome in the future.

3.1.7 Hazard Identification

This is the method of formally collecting, recording, and analysing potential hazards. This data can then be used to take any corrective action and distribute feedback to the relevant places. It needs to be a perpetual process.

Some suggested methods to assist with this are as follows:

- Brainstorming—group meetings
- Data analysis from previous accidents/incidents
- Internal or external mandatory and voluntary incident-reporting schemes
- Internal or external audits/safety assessments
- Safety information from external sources such as media, similar organisations, AIIB, CAA

3.1.8 Hazard Log

The hazard log is used to keep a detailed and accurate record of all identified events that could contribute to an accident or incident. We can use this data to analyse and keep track of what is causing occurrences and use this to develop mitigation techniques and strategies to try and prevent re-occurrences.

All identified safety hazards, Risk Assessments, and subsequent corrective actions need to be clearly documented and accessible to anyone within the organisation.

The creation of a hazard/risk register is an acceptable process which must include:

- Giving individual hazards a specific identifier
- A general overview of all associated risks
- The results of any Risk Assessment
- Any current mitigation procedures in place
- Any further or future mitigation procedures that are required
- Re-assessment of mitigation procedure to confirm the change in risk level with the desired outcome

Risk Assessment is essentially a process that enables the identification of hazards and risks, calculating the probability and severity of the event occurring, to determine the appropriate control measures.

Throughout all aspects of operations, hazards and risks must be identified and assessed and suitable control measures implemented to reduce the risk to a level deemed ALARP. Risk management will commence during operational planning and throughout the operation. All flight crew must adhere to the control measures detailed within the Risk Assessment, and the RP is responsible for monitoring the implementation of these.

The production of a Risk Assessment for each UAS Operation is a mandatory requirement, and the format of such can be found on the Operational Planning form within the Operations Manual. Risks are assessed against the Risk Matrix detailed on the Risk Assessment document.

LIKELIHOOD OF OCCURRENCE		
Qualitative Definition	Meaning	Value
Frequent	Likely to occur many times (has occurred frequently)	5
Occasional	Likely to occur sometimes (has occurred infrequently)	4
Remote	Unlikely to occur but possible (has occurred rarely)	3
Improbable	Very unlikely to occur (not known to have occurred)	2
Extremely Improbable	Almost inconceivable that event will occur	1

FIGURE 3.3 Risk Likelihood table.

3.1.9 Risk Likelihood

Risk Likelihood is the process where an organisation can estimate what the statistical probability of an event occurring is. They can then implement mitigating measures, via several methods, in order to reduce the risk to an ALARP (As Low As Reasonably Practicable) level.

Next, we have some questions that assist with looking at Risk Likelihood.

- Is there a history of similar occurrences (either in your own organisation or other similar organisations)?
- Is this an isolated occurrence?
- What other aircraft, equipment, or components of the same type may have similar defects?
- How many people are involved?
- How frequent is the activity?

3.1.10 Risk Severity

Risk Severity refers to a method we can use to quantify what the impact of an occurrence happening is, e.g. financial, organisational, or reputational impact.

We do this by referring to a risk matrix scoring system. This can be factored along with other risk data in order to make an objective assessment on what needs to be done to mitigate the risk. When doing this, we must take into account the concept of ALARP.

The following questions may help to assess the severity of risk:

- Would lives be lost (employees, bystanders)?
- What is the likely extent of property or financial damage?
- What is the likelihood of an impact to the environment (contamination or disruption to natural habitat)?
- What are the likely commercial or media publicity implications?
- Would there be a loss of reputation?

SEVERITY OF OCCURRENCE		
Aviation Definition	Meaning	Value
Catastrophic	Results in an accident, death, or loss of equipment.	5
Hazardous	Serious injury or major equipment damage	4
Major	Serious accident or injury	3
Minor	Results in a minor incident	2
Negligible	Nuisance of little consequence	1

FIGURE 3.4 Risk Severity table.

3.1.11 Risk Tolerability

Risk tolerability is the simplified method of assessing the overall impact of a risk on the organisation after it has been documented. It allows for a much simpler decision to be made by the Accountable Manager or UAS Operations Manager regarding the possible outcome of any documented hazards.

The assessment of the tolerability of risk may be estimated by the use of a Risk Tolerability Matrix and classified as:

- **Unacceptable**: The operation should be stopped immediately and not take place. Major mitigation is necessary to reduce the severity or reduce the likelihood of it occurring.
- **Review:** The current operation raises some concerns, and methods such as low as reasonably practical (ALARP) should be implemented under the authority and discretion of the Accountable Manager.
- **Acceptable**: If the risk is considered acceptable, then it should be assessed if it can be reduced further by any appropriate means.

Risk Tolerability Matrix

RISK SEVERITY					
Risk Likelihood	**Catastrophic 5**	**Hazardous 4**	**Major 3**	**Minor 2**	**Negligible 1**
Frequent 5	Unacceptable	Unacceptable	Unacceptable	Review	Review
Occasional 4	Unacceptable	Unacceptable	Review	Review	Review
Remote 3	Unacceptable	Review	Review	Review	Acceptable
Improbable 2	Review	Review	Review	Acceptable	Acceptable
Extremely Improbable 1	Review	Acceptable	Acceptable	Acceptable	Acceptable

FIGURE 3.5 Risk Matrix table.

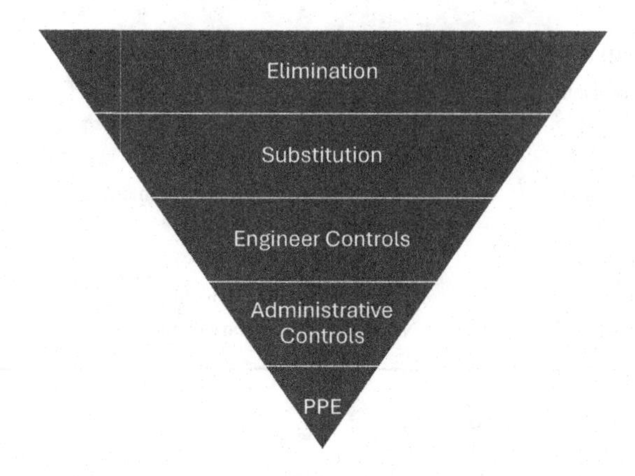

FIGURE 3.6 Industry standard hierarchy of control.

Using the Risk Tolerability Matrix is a simple process. We check what the allocated Risk Likelihood is against the Risk Severity using the Matrix table. The result will give us a simplified tolerability that can be used to further mitigate any hazards.

3.1.12 Hierarchy of Control

The hierarchy of control is used in many industries and has become the standard for choosing control measures bringing risks to a level deemed As Low As Reasonably Practicable (ALARP). Controls must be implemented in the order as shown in Figure 3.6 where possible, because controls higher up the hierarchy have been proven to be more effective than those below. It is not acceptable to jump straight to the easiest option to implement. As a last resort, if the risk can be ALARP through the adoption of the hierarchy of control, Personal Protective Equipment (PPE) must be issued to the individual(s) at risk.

Controls must be implemented in the order shown below where possible, because controls higher up the hierarchy have been proven to be more effective than those below.

3.1.13 Assessing if a Safety Risk Is ALARP (As Low As Reasonably Practicable)

A risk is considered to be ALARP when any further reduction in either the likelihood or severity would not be possible without expending a disproportionate number of resources. It is important to recognise that the assessment of ALARP is a process different from the assessment of tolerability. This means, even when a risk has been assessed as tolerable, it must also be assessed as ALARP.

3.1.14 Mitigating Risks

Mitigations are the measures that we introduce to reduce the safety risk level. This is achieved by reducing either the likelihood or the severity of the risk. Mitigations are needed when a risk has been assessed as either not tolerable or not ALARP. Mitigations may either be operational or technical in nature, and the robustness of the mitigation is an expression of the confidence that it can, and will, reduce a risk's level.

3.1.15 Identifying a Hazard Case Study Scenario

Small Aircraft Maintenance Ltd is an organisation performing maintenance on a fixed-wing aircraft below 2,739 kg. During routine maintenance, an engineer found wire locking missing on the attachment bolt of a connecting rod for the aileron system. The engineer reported the finding on a Safety Report Form. An incident investigation was carried out which concluded that the error occurred during an aircraft maintenance check some weeks before. The error occurred due to time pressure because the aircraft maintenance check was running late and the aircraft was required for flight. Since then, the aircraft had completed a number of flights.

3.1.16 Associated Risk

During the next Flight Safety Meeting, the Safety Committee discussed the reported hazard. The committee consisted of:

- The Accountable Manager
- Chief Engineer
- Chief Pilot
- A private Pilot
- An engineer

The Safety Committee defined the hazard as incorrect maintenance action and identified a possible risk of the connecting rod becoming detached causing a loss of control of the aircraft. The following demonstrates the reporting form that was completed during the committee meeting.

Part A: To be completed by the person identifying the event or hazard.

Date of event		Local time	
Location			
Name of reporter		Section Department Organisation	

Please fully describe the event of identified hazard include suggestions on how to prevent similar occurrences

What is the likelihood of such an event or similar happening again?				
Extremely improbable			Frequent	
1	2	3	4	5
What do you consider could be the worst possible occurrence if this event did happen again?				
Negligible			Catastrophic	
1	2	3	4	5

Part B: To be completed by the Flight Safety Officer

The report has been recorded on the company database			
Report reference No		Date	
Signature			
Name			

FIGURE 3.7 Hazard report.

Date	19 April 2010		
Owner	A Smith	Contact Number	07712345678
Participants	F Bloggs (Accountable Manager), T Jones (Chief Pilot), A Smith (Chief Engineer), B Edwards (Mechanic), B Simms (Pilot)		
Reported by	B Edwards	Contact Number	07798745612
Date Reported	2 March 2010		
MORS Filed (yes/No)	Yes		
Follow up review date	19 July 2010		

Identified Hazard	Associated Risk (Consequences)	Existing Mitigation (measures in place)	Current Level of Risk	Further Mitigation Measures	Revised Level of Risk	Action By and When
Incorrect maintenance action Wire locking missing from aileron system connecting rod	Connecting rod detaches causing loss of control of aircraft	Aircraft service manual instruction to wire lock the connecting rod bolt. Duplicate engineering inspection required	Severity 5 Likelihood 3 Unacceptable	Reiterate adherence to Aircraft Service Manual and independent inspections. Introduction of staged worksheets for breakdowns. Implementation of a Maintenance Error Management System (MEMS)	Severity 5 Likelihood 2 Review	A Smith, July 2010

FIGURE 3.8 Safety Officer and the Safety Committee report.

Over the following page, we will break down the cycle they went through, to mitigate the hazard to ALARP.

3.2 EXAMPLE SAFETY REPORTING FORM

The form given here is what the Safety Committee would have completed during the meeting.

Part C: To be completed by the Safety Committee

Rate the likelihood of the event occurring or reoccurring				
Extremely Improbable			Frequent	
1	2	3	4	5
Rate worst case consequences				
Negligible			Catastrophic	
1	2	3	4	5
What action or actions are required to eliminate, mitigate, or control the hazards to As Low As Reasonably Practicable (ALARP)?				
Resources required				
Person responsible for action				
Agreed and accepted by:	Safety Officer		Date:	
	Accountable Manager			
	UAS Operations Manager			
	Responsible Manager			
Appropriate feedback given to staff by Flight Safety Officer/Accountable Manager/UAS Operations Manager				
Signed			Date	
Follow-up action required				
Responsible Person				
Date to complete by:				
Date Hazard log updated:				

FIGURE 3.9 Safety Officer and the Safety Committee report continued.

3.2.1 Further Mitigation Measures

As the hazards should always be mitigated to an ALARP level, we usually need to add further mitigation measures in order to reduce the likelihood or severity of the hazard. If the Risk Level has an acceptable outcome, it can be left there. However, if the outcome is that the risk was in the unacceptable category, major mitigation measures were required to reduce the level of risk to As Low As Reasonably Practicable (ALARP). The Safety Committee identified a number of further mitigation measures you can see in the table given next.

Identified Hazard	Associated Risk (Consequences)	Existing mitigation (Measures in Place)	Current level of Risk	Further Mitigation Measures
Incorrect maintenance action. Wire locking missing from aileron system connecting rod.	Connecting rod detaches causing loss of control of aircraft	Aircraft Service Manual Instruction to wire lock the connecting rod bolt. Duplicate engineering inspection required	Severity 5 Likelihood 3 Unacceptable	Reiterate adherence to Aircraft Service Manual and independent inspections. Introduction of staged worksheets for breakdowns. Implementation of a Maintenance Error Management System (MEMS)

FIGURE 3.10 Further mitigating actions by the Safety Officer and the Safety Committee.

3.2.2 Revision Level of Risk

The risk was reassessed in terms of severity and likelihood, taking into account the further:

- Mitigation measures introduced
- With the new measures in place, although the severity remained the same, the conclusion was that the likelihood of the risk occurring was now improbable (2)

Using the Risk Assessment matrix, with the Risk Severity determined to be catastrophic (5) and the likelihood of occurrence determined to be improbable (2),

Identify Hazard	Associated Risk (Consequences)	Existing Mitigation (Measures in Place)	Current Level of Risk	Further Mitigation Measures	Revised Level of Risk	Action By Whom and When
Incorrect maintenance action. Wire lock missing from aileron system connecting rod	Connecting rod detaches causing loss of aircraft control	Aircraft Service Manual instructions to wire lock the connecting rod bolt. Duplicate engineering inspection required.	Severity 5 Likelihood 3 Unacceptable	Reiterate adherence to Aircraft Service Manual and independent inspections. Introduction of staged worksheets for breakdowns. Implementation of Maintenance Error Management System (MEMS)	Severity 5 Likelihood 2 Review	A Smith July 2010

FIGURE 3.11 Review of corrective actions.

the risk was now classified in the review category. Although the Safety Committee agrees that the risk had been mitigated to As Low As Reasonably Practicable, it was accepted that a level of risk still remained.

3.2.3 Scenario Outcomes

As part of Small Aircraft Maintenance Ltd.'s safety assurance, the Safety Committee decided that all the critical flight control systems of the aircraft they maintained should be inspected to see if there were any similar defects.

3.2.4 Safety Training and Communication

The Safety Committee also produced a safety bulletin to remind engineers of the importance of wire locking critical flight control systems and introduced a training session to highlight the new procedures for staged worksheets and the introduction of the Maintenance Error Management System (MEMS).

3.2.5 Hazard Identification

3.2.5.1 Before Flight

There are a range of components available to assist with identifying any hazards present before a flight takes place.

3.2.5.2 Desktop Study (Operational Planning)

The obvious component is during the planning stage, where we are looking initially at the proposed task and its feasibility. A thorough plan should identify anything present in the area of the flight which may present a hazard to the operation of the UAS.

3.2.5.3 Aircraft User Manuals and Flight Reference Cards (FRCs)

User manuals and FRCs should inform the operator of any Failure Points or Single Points of Failure (SPOF) present in the system. The operator should be able to create a generic Risk Assessment for the specific aircraft.

3.2.5.4 Flight Safety Meetings

During Flight Safety Meetings, any issues that have occurred with aircraft or operations should be discussed. These discussions have provided critical information on how the hazard occurred and recommended mitigations for future operations.

3.2.5.5 Pilots-to-See (PTS)/Stop Press

Usually resulting from a Flight Safety Meeting, a well-developed Safety Management System should employ the use of a PTS or Stop Press system. These are used to disseminate operationally critical information around the organisation and provide a wealth of information on any known hazards, as well as recommended mitigations. Next, we have an example of a Risk Assessment that was completed to show the hazards that may occur when managing a UAS Programme.

3.2.6 Risk Assessments

Risk Assessments form a crucial part of any Safety Management System. They create company-wide awareness of hazards and their required mitigations and will provide an organisation with value in a range of areas.

By promoting the use and dissemination of Risk Assessments, it will lead to an increase of awareness throughout an organisation.

Risk Assessments will highlight specific users or groups that may need to be aware of hazards related to their job or operation.

Risk Assessments are a legal requirement of HSE regulations (PUWER). Additionally, as they are a CAA requirement detailed within a company Operations Manual, any insurance or Operational Authorisation being used will only be valid if a suitable Risk Assessment has been carried out.

Risk Assessments detail the standard mitigations and any further mitigation requirements for controlling the hazard to an ALARP level. This could affect what equipment a Remote Pilot must bring to an operational task or when working in a specific environment, for example, PPE requirements, warning signs, or extra UA Observers.

Hazard	Risk	Existing Mitigation (control measures)	Risk Level (P × S = L)	Further Mitigation (Control Measures)	Risk Level (P × S = L)	ALARP (Y/N)	Risk Owner
			P S L		P S L		
Missing or Unserviceable Safety Equipment	Equipment not available or unserviceable when needed	RP to check prior to deploying on an operation	3 4 2	Monthly serviceability checks of all ancillary equipment. Recorded in equipment log	1 4 4	Yes	Initial or signature of risk owner
RP out of currency	Operations unable to proceed if breach of Operations Authorisation not noticed	RP responsible for monitoring own currency	3 4 12	Ops Manager or Accountable Manager to carry out Quarterly Logbook checks.	1 4 4	Yes	Initial or signature of risk owner
Aircraft unserviceable	Catastrophic aircraft failure during mission or postponing of operation	RP to carry out pre-flight checks IAW Operations Manual and Manufacturers guidelines	2 5 10	Monthly Maintenance checks to be carried out IAW Manufacturer Guidelines. Recorded in Aircraft Maintenance Log.	1 5 5	Yes	Initial or signature of risk owner
Operations not being carried out in accordance with Operations Manual or Operations Authorisation	Breach of Op Auth. Serious incident during operational flight	RP to ensure all operations are carried out IAW Ops Man and Operations Authorisation	3 4 12	Spot checks carried out by appropriate person. Updated documentation disseminated to all RPs. RPs to sign when read and understood updates.	1 4 4	Yes	Initial or signature of risk owner
RP not fit to fly	Pilot incapacitation incident during operation	RP to carry out IMSAFE checklist prior to operation	2 5 10	Spot checks carried out by appropriate person. RP to sign a Fit-to-Fly register before flight	1 5 5	Yes	Initial or signature of risk owner

FIGURE 3.12 Hazard log.

Ultimately, Risk Assessment helps to promote a healthy, safety-conscious attitude within an organisation. This will lead to employees feeling safer when working and should promote the willingness to drive safely as a priority. The end results will be safe, efficient working practices and could also present a reputational or financial benefit.

4 Safety Assurance and Compliance

4.1 MANAGEMENT SYSTEMS OVERVIEW

In this chapter, we will investigate the different methods of ensuring that the processes described in the Safety Management System (SMS) can be policed and monitored. This is to ensure that legal compliance and the effectiveness of a robust SMS are being maintained.

A distinct advantage in maintaining the disciplined structures we have looked at previously is allowing an organisation to keep track of events and use the data to operate a safer and more transparent business.

From a legal perspective, having a well-structured compliance programme allows internal company auditors to complete compliance checks with ease and shows any regulating authorities, such as the CAA, that the organisation is taking their compliance seriously.

4.2 SAFETY PERFORMANCE MONITORING AND MEASURING

To maintain regulatory compliance and monitor organisation performance, all safety data needs to be measured and documented. This can be achieved by establishing specific methods of measuring often known as Safety Performance Indicators (SPIs) or Key Performance Indicators (KPIs).

SPI's may be based upon

- Events
- Frequency and attendance at safety meetings
- Reporting levels

Safety reports may include their categorisation including:

- Business area of reporter
- Type of event
- Type of aircraft
- Type of equipment

The table given next gives some examples of Performance Indicators, their objectives, and a record of their achievement. These are specific to Operations and Operators.

DOI: 10.1201/9781032620220-5

Performance Indicator	Objectives	Performance											
		1	2	3	4	5	6	7	8	9	10	11	12
		Qtr 1			Qtr 2			Qtr 3			Qtr 4		
Number of major risks incidents (as defined in SMM)	1 or less												
Number of MORS	3 or less												
Number of internal audits	4												
Number of audit findings per audit	2 or less												
Number of Safety Committee meetings	6												
Safety Committee attendance of key personnel	Minimum 80%												
Number of ERP drills	1												
Number of Safety/ Hazard reports	20 or more												
Number of Safety newsletters issued	2												
Number of formal Risk Assessments	5 or more												
Number of safety surveys	1												
Number of airworthiness incidents as defined in SMM	1												
Number of flights flown with MEL restrictions	3 or less												

FIGURE 4.1 Key or specific Performance Indicators.

Note: 1. Safety Objects specific to an operator.
2. The suggested objectives are an example only. Organisations should set objectives that are relevant to their particular type of operation.

4.3 SOURCES OF SAFETY DATA

Some sources of Safety Data that can be used to measure and monitor Performance Indicators are given as follows:

- Hazard and incident reports
- Warranty claims

- Customer complaints
- MORS
- Customer/contractor surveys
- Safety surveys
- Safety audit findings
- Air Accident Investigation Branch (AAIB) reports
- Manufacturer known issue lists
- Open-source platforms (forums/social media)

4.4 CHANGE MANAGEMENT

In any dynamic industry, changes occur frequently. It is therefore essential that a specific and structured process is employed to identify the impact any such changes might have upon the operation.

The Change Management process should be a clearly defined method of identifying and assessing the safety and commercial implications of such changes, rather than a random method of assessment.

Some examples include:

- Introduction of new equipment
- Changes to facilities
- Changes to scope of work
- New operational tasks
- New contracted services
- New regulatory procedures
- Changes of key staff
- Are existing procedures and documentation adequate?
- Have all staff received appropriate and sufficient training?

All changes should be assessed using the organisations standard Risk Assessment process to determine any impact upon current and future activities.

4.5 INCIDENT MANAGEMENT

Another process which should be well rehearsed and routinely practised is that of Incident Management. Any effective Safety Management System should allow learning from events in order to react appropriately and implement any required operational changes within a safe timeframe.

This may be achieved by investigating all accidents and incidents at a level appropriate to their significance, including:

- What happened?
- When?
- Where?
- How?
- Who was involved?

The number one priority is to discover why and prevent any re-occurrence. This is also the foundation of all modern Human Factors principles—to avoid speculation by establishing facts. A safety committee or Safety Officer should review all findings and then recommend. improvements. These findings should be shared with all appropriate parties, both inside and outside of the organisation.

4.6 SAFETY ASSURANCE AND COMPLIANCE MONITORING OF THE SMS

An integral part of a Safety Management System is the establishment of Compliance Monitoring (Quality Management). This requires independent, objective, and continuous assessment. It should include:

- A regular review of how the organisation complies with operational and legislative requirements of the SMS
- Transparent verification process to confirm that mitigations and controls to identify hazards are effective and robust
- An assessment of the effectiveness of Safety Management Systems processes and procedures
- How any processes and procedures are implemented and practised
- Regular internal and external audits verify that any problems impacting upon safety are identified and corrected within an appropriate timeline

This is a closed-loop process which requires the Accountable Manager to be aware of what is revealed through investigations. There may be legislative legal requirements dictating the structure and operation of all compliance management. In Figure 4.2 we have an example of the Safety Management System Continuous Improvement Cycle.

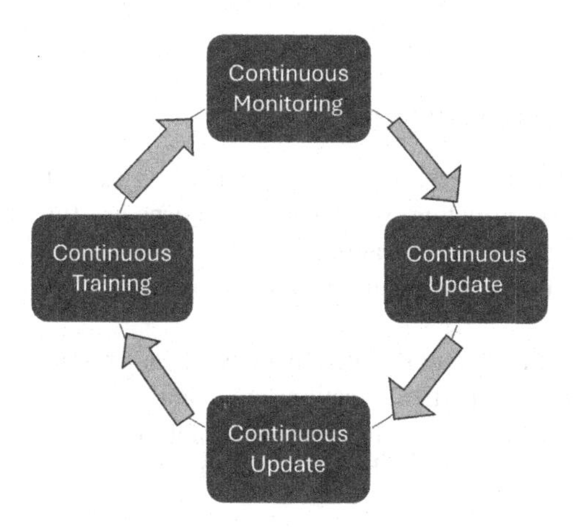

FIGURE 4.2 The Safety Cycle.

4.7 SAFETY MANAGEMENT STRATEGIES

To reinforce the principle of continuous development, it is imperative to analyse data in the context of a strategy which is a combination of the following:

- **Reactive**—responding to events (accidents and incidents) which have already occurred.
- **Proactive**—actively identifying hazards by using the organisations processes (e.g. Q Pulse Reporting System).
- **Predictive**—analysis of system processes and environment to identify potential future problems.

4.8 SAFETY TRAINING AND EDUCATION

In line with the concept of Continuous Training from the Safety Cycle, there should be a programme of regular Safety Training and Education. Everyone in your organisation has responsibility for safety. To ensure the success, all staff members should receive regular training and ongoing assessment of understanding and competence. The training should include subjects such as:

- Safety policy
- Reporting procedures
- Safety responsibilities
- Periodic documented refresher training
- How staff at *all* levels can contribute to the effectivity of the SMS?

When completing any internal training, compliance can be achieved by recording all staff training. During training, staff should be actively encouraged to identify and report any hazards. This can be facilitated though providing the awareness of specific safety hazards associated with operational duties.

These training sessions are an ideal environment to actively disseminate any lessons learnt during safety investigations.

4.9 SAFETY COMMUNICATION

It is essential that all persons working for or with your organisation as full-time, part-time employees, subcontractors, volunteers, or interns are fully aware of the SMS and all safety aspects of your operations.

This is achieved by effective communication, including:

- Safety critical information related to analysed hazards and assessed risks
- Understanding why particular actions are taken
- Why new safety procedures may have been introduced
- Why safety procedures may have been changed

- Regular safety meetings, newsletters, notice boards, and online resources should be used to publicise and disseminate all information to promote awareness and importance of safety
- All safety data should be shared and exchanged with other similar organisations

4.10 GAP ANALYSIS

Operating with a fully functioning effective Safety Management System is an ongoing dynamic process.

The principles of continuous monitoring, review, update, and training as set out in the Safety Cycle form the foundation of your SMS.

To effectively ensure this evolving process, it requires a routine identification of what components are in place and what new additions will need to be implemented or strengthened to ensure maximum performance.

Using a Gap Analysis process will assist in assessing any needs. A comprehensive Gap Analysis may be achieved by applying an organisation review using a routine internal audit process. The preparation and understanding of what external auditors are looking for give an excellent impression and often assist with a more favourable assessment due to the confirmation that your organisation has a professional attitude towards safety. An Implementation Plan The best way to effectively develop your SMS into a robust system is to follow a planned course of evolution over time. A logical, structured, and realistic road map is desirable.

The Gap Analysis process will facilitate this development. Some points to consider when implementing are as follows.

- A mature Safety Management System takes time to evolve.
- All staff members should have the opportunity to contribute to this development.
- The CAA surveyors or inspectors will gladly offer any advice.

In Chapter 22, we have provided several "Typical CAA Audit questions", and these form an excellent tool to help develop a comprehensive Gap Analysis and Compliance programme within your organisation.

5 Safety Promotion

5.1 SAFETY PROMOTION AND COMMUNICATION

It is essential to routinely and regularly promote and maintain a culture and ethos for safe UAS operations. Many short-sighted organisations do not understand the "hidden" value of promoting a transparent operational structure which highlights the best practice business structures and operational processes. Often, a company with a clearly recognisable and functional SMS has a financially competitive edge compared to many of its business competitors.

5.1.1 How Can We Promote Safety?

One of the most ignored but successful methods to promote safety is through a company culture which values and provides continuous training to facilitate Continued Professional Development for all stakeholders and employees. Safety Promotion begins with management support with leading by good example as the most basic principle. Your organisation could consider launching a series of safety campaigns. This could include, but is not limited to:

5.1.1.1 Safety Policies and Values

Safety policies and values should be published to describe the organisation values and commitment to employee welfare and expectations and commitment to external customers. They should clearly define acceptable and unacceptable behaviour with a clear statement of how this will be consistently and transparently enforced.

Rules and regulations should be minimised where possible and clearly defined in clear and simple language. These rules should only be the necessary requirements defined using common sense. All employees should receive robust induction training and orientation upon joining and regular refresher and updated training with appropriate rewards.

5.1.1.2 Keynote Briefings

A tried and tested very successful method to promote and maintain safety awareness is to precede each gathering or meeting with a few minutes devoted to a review of the monthly SMS topic. This could be a short video or a Q&A session. One logical method is to assign the monthly topic with reference to the most recognised "12 Pillars of SMS" as advocated by ICAO:

January	**Management Commitment**
February	**Safety Accountability and Responsibilities**
March	**Appointment of Key Safety Personnel**
April	**Coordination of Emergency Response Planning**

DOI: 10.1201/9781032620220-6

May	**SMS Documentation**
June	**Hazard Identification**
July	**Safety Risk Assessment and Mitigation**
August	**Safety Performance Monitoring and Measurement**
September	**Management of Change**
October	**Continuous Improvement of the SMS**
November	**Training and Education**
December	**Safety Communication**

5.1.1.3 Promotional Literature (Bulletins, Leaflets, and Posters)

The briefing just described can be supplemented with an appropriate handout reinforcing the message. Hardcopy or online newsletters should be regularly disseminated to keep all organisation members updated with information concerning safety issues, developments, and future plans keeping everyone up to date and in the "Company Loop". This keeps staff and management sharing common goals and responsibilities to foster a spirit of good teamwork. Posters with sensible organisation aims targets and motivational messages can be very powerful if well written.

Regular Flight Safety Bulletins should be a feature of any promotional literature.

5.1.1.4 Promotional Videos, Podcasts, and Apps

With the current advances in technology, it is easy to develop and produce podcasts and videos which may be more convenient for employees on the move to supplement or even replace the promotional literature already described. Another advantage of this approach is that it facilitates the publishing of internal company surveys or questionnaires to allow the exchange and gathering of important data.

5.1.1.5 Case Studies

Case studies are an excellent and interactive method to investigate accidents, incidents, and occurrences from within your own organisation or more frequently from external sources. In group sessions, attendees seem to particularly enjoy the process of analysis and comparison to their own experiences and make any connections to similar events which may have occurred in their own organisation.

Later in this publication, we will be exploring several case studies in much more depth, suggesting structures and processes which make the task of case study analysis more understandable and logical.

5.1.1.6 Safety Awareness Events

It is very productive to attend as many Safety Awareness events as possible. Generally, you will find that many other contributors will have in the past or will do in the future experience very similar or even identical occurrences during their routine activities. By collaborating and sharing these experiences with your own colleagues and even more so by inviting third-party contributions, a massive amount of safety information can be gathered. In many cases, this knowledge can prevent the repetition of the same mistakes as experienced by others. With regard to an organisation developing their own safety awareness events, the transparency of process can often yield hidden commercial benefits and demonstrate that your own organisation's commitment to safety is an important priority.

5.1.1.7 Industry Roadshows

As suggested earlier, such events are the perfect way to showcase your organisation and its commitment to quality.

5.1.1.8 Quick Reference Guides

Producing appropriate operational checklists and quick reference guides is an excellent method of ensuring that operators adhere to Standard Operating Procedures (SOPs) which are developed to mandate that best safety and efficiency processes are followed.

5.1.1.9 Wellness Programme

Companies should provide programmes to help employees adopt a healthier lifestyle to improve their physical and mental health with support involving diet, exercise, stress management, and sharing/team-building activities.

5.1.1.10 Engagement with Relevant External Safety Agencies

Connecting, utilising, sharing, and developing a relationship where possible with the many helpful Safety Agencies to learn and discuss is invaluable. A few examples are listed in the next section, but a much more comprehensive listing with contact details may be found in the Appendices.

5.2 DEVELOPING YOUR OWN SMS

This book has been written to highlight best practice in SMS as applied in the manned aviation context and may well contain much detail which may not be relevant to your own organisation. However, as your organisation grows and the RPAS Industry becomes more sophisticated and regulated, you will notice the relevance more acutely. An SMS is a living, breathing system which needs to be constantly updated and reviewed, but most of all it needs to be your SMS. It needs to be appropriate to the specific requirements of your own organisation.

5.3 KNOWLEDGE CHECK

The following questions will give the reader the opportunity to check their progress and understanding of the previous chapter content prior to a review of learning outcomes.

1. What is the ICAO definition of a Safety Management System?
2. Where can the definition be located?
3. Give four reasons why adopting an SMS is necessary.
4. What is meant by ALARP?
5. List the four functional components of an SMS.

6. Give five of the requirements of a Safety Policy.
7. List two of the main components of hazard management and Risk Analysis.
8. Briefly describe the functions of Quality Management.
9. Why do you need to promote safety?
10. Explain three factors considered when conducting a Risk Assessment.
11. What is an ERP?
12. Name the five components of the Safety Cycle.
13. What is a Risk Assessment?
14. What is Risk Likelihood?
15. What is Risk Tolerance?
16. What is Risk Severity?
17. What is the Risk Tolerance Matrix?
18. What are the four elements of the SMS Operational Cycle?
19. List the elements contained in the Hierarchy of Control diagram.
20. Why is it important to formally promote safety?

Before continuing to the next chapter, it is worth reviewing whether or not you believe you have achieved and fully understood the Learning Outcomes from Chapter 1. If you try to achieve at least 16 out of 20 correct answers (80%) from the answer list contained in the Appendices, that is normally the standard required for pass level in aviation.

LO 1: Understand the basic principles of Safety Management Systems.
LO 2: Describe how Safety Management principles can be applied to RPAS Operation.
LO 3: Understand the importance of hazard management for RPAS Operations.
LO 4: Analyse the four main components of Safety Management Systems.

Part Two

Human Factors (HF), Crew Resource Management (CRM), Team Resource Management (TRM)

FIGURE PII.1 Time lapse photograph of a night-launched drone.

DOI: 10.1201/9781032620220-7

6 Human Factors (HF), Crew Resource Management (CRM), Team Resource Management (TRM)

6.1 LEARNING OUTCOMES

By the end of the lesson, students will be able to:

LO 1: Analyse what factors influence various human behaviours.
LO 2: Investigate appropriate processes to understand, identify, and mitigate those behaviours which affect operational safety.
LO 3: Describe the importance of having a structure allowing the classification and grading of non-technical behaviours.
LO 4: Identify and explain why subjects are described as generic behaviours and job specific.

6.2 OVERVIEW

Aviation research confirms that 70 to 80% of all accidents/incidents are caused by issues arising from human Factors or non-Technical root causes. Further analysis over the past 50 or so years demonstrates the evolution of thinking. In the early days when technical reliability was responsible for most accidents/incidents, researchers managed to mitigate many of these problems by refining the technical solutions available to them. Some examples are engine reliability on piston to gas turbine to pure jet, improved construction materials, advances in communication and information systems such as RADAR, cockpit voice recorders, black box telemetry data collection, and warning systems such as Ground Proximity Warning Systems, Radar Altimetry, and Secondary Surveillance Radar (SSR). As all these systems developed into more sophisticated and reliable examples, it became obvious that still the human factor was the most prevalent influence of occurrences.

DOI: 10.1201/9781032620220-8

For a comprehensive investigation of this data see the following:
A Statistical Analysis of Commercial Aviation Accidents 2023.

After a lifetime in the Aviation Industry, it is obvious that many of the problems faced by individuals and organisations come from different departments operating with a "silo mentality" where they are completely unaware of many of the issues faced by their employees and colleagues and often just do not care.

Attitudes to human performance are so often not appreciated or understood, and many organisations fail to conduct appropriate initial and refresher training. The Aviation Industry finally woke up and realised that Human Factors (HF) training needed to be a mandatory requirement, and so it became a legal requirement to conduct suitable training first for Pilots and then engineers. When I delivered CRM training to Pilots at a major helicopter company, sometimes, in the next classroom, there would be another class of engineers doing almost exactly the same training syllabus but restricted to each respective discipline. This attitude obviously misses the point and a very special opportunity.

It soon became obvious that everyone in the organisation would benefit from similar training but only if all were sitting together in the same room. I tried with much opposition from senior management to integrate different departments and disciplines to come together for shared training and experience. I invited several external participants from universities and associated industries including customers. This was very well received, and feedback suggested that this was the future of Human Factors training.

The main obstacle was that Senior Management would not support the concept since they said it was too expensive and time consuming and not a regulatory requirement except for Pilots and engineers.

However, to a certain extent while much of the Aviation Industry has remained a little complacent, other disciplines are evolving those basic principles into more forward-looking concepts and moving in the direction of Team Resource Management (TRM). Practically, I have come to the conclusion that one of the best solutions is to classify and conduct training into two phases, **GENERIC HF** behaviours applicable to all organisation members and **JOB SPECIFIC** as in the case of commercial aviation to Pilots, engineers etc.

The *Generic* class would include staff and guests from all company disciplines, while the **JOB SPECIFIC** should focus upon the specific technicalities and aspects of each specific discipline.

The following chapter highlights the examples of the subjects which received so much successful feedback.

Obviously, staff would attend both examples of the training geared to their legal compliance and best practice requirements.

The treatment of the subject in this publication is offered as a very basic subject introduction/overview to one of the most fascinating subjects which is still finding its place in many different industries.

I have focused upon what I consider to be the most practical take away points which are relevant to Pilots and Engineers within the time available allocated to training by most organisations. Sadly, this often just provides an overview, and in my own opinion should be covered in much more depth which one day may be the case.

I have provided several links to short videos and documentation, which demonstrate the principles.

If it stimulates your interest and imagination which I hope it does, I would suggest you to read at least one but preferably both titles considered to be the definitive "Bibles" on the subject:

- ***Safety at the Sharp End: A Guide to Non-Technical Skills* by Rhona Flin, Paul O'Connor**
- ***Crew Resource Management* by Barbara G. Kanki (Editor), José Anca (Editor), Thomas R Chidester**

To put Human Factors into perspective, I recommend everyone to watch the video, "Just a Routine Operation" by **Captain Martin Bromiley**

6.2.1 What Are Human Factors?

When learning to perform any task, we need to be trained by some appropriate method. This could be observation, trial and error, mimicking, or formal instruction. Through these methods, we gain a basic understanding. Once the concept is understood, we must now practise it. In time, we develop proficiency, and by repetition, we gain experience. With enough experience, the performance of such a task can be considered a skill. There are many factors influencing the success of this process, and the most significant may often be the attitude or behaviour of the student. In studying Human Factors, we can begin to understand how different individuals learn with different techniques, at different rates, and how the retention of such knowledge can vary. As humans, we are all complex and wired differently. We react to stimuli in many different ways. Human Factors have largely been pioneered by the Aviation Industry, and the study of human behaviour has opened up vast new understandings of how we can work together more safely and effectively by just taking the time to consider how individuals and groups interact, what motivates us to do what we do, and act as we do. If we can understand, recognise, and respect the drivers that stimulate what makes us similar, as well as different, it follows that we can improve teamwork and productivity. This fascinating and emerging science, which is a branch of Applied Industrial Psychology, crosses over into several other studies such as Wellness, Economics, and Project Management. Known variously as Human Factors (HF), Crew Resource Management (CRM), and Team Resource Management (TRM), this subject tries to understand, research, and explain the complexity of the human and how we can best work together.

6.2.2 Introduction to Human Factors

The UAS industry is expanding and evolving at a rapid pace. As it becomes more sophisticated and technical, the regulatory authorities will ultimately impose more restrictions and rules, which will be a pre-condition of operating all UAS. The status and respect that operating UAS now commands, along with the recognition from regulatory authorities, as more complex operations with larger and more complex drones take place, have obvious consequences. UAS Operations are now rightly considered to be as important to the aviation sector as manned operations. There is no doubt that over the next few years, the regulation requirements will begin to align more with these other sectors.

6.2.3 Crew Resource Management (CRM) Definition

Crew Resource Management (CRM) is a management system which makes an optimum use of all available company resources' equipment, procedures, and people to promote safety and enhance flight operations by reducing error, avoiding stress, and increasing efficiency. CRM is concerned not so much with the technical knowledge and skills required to fly and operate an aircraft but rather with the cognitive and interpersonal skills needed to manage the flight with an organised aviation system. In this context, cognitive skills are defined as the mental processes used for gaining and maintaining situational awareness, for solving problems, and making decisions. Interpersonal skills are regarded as communications and a range of behavioural activities associated with teamwork. These skills often overlap with each other, and they also overlap with the required technical skills.

6.2.4 Statistics

Aviation is a complex, safety critical environment, in which an unsuitable action can lead to major consequences. Human error is the cause of approximately 80% of all aviation accidents, and thus CRM is a hugely important part of the mitigations available to reduce the probability of errors and thereby improve Flight Safety.

6.2.5 Why Do Human Factors Matter?

One of the considerations in maintaining the highest degree of safety is related to the potential costs of any accidents or incidents.

6.2.6 Human

Implications may be:

- Tragic loss of life
- Injury

6.2.7 Public

Implications may be:

- Perception
- Reputation
- Bad publicity
- Loss of existing customer and potential new customer confidence

6.2.8 Financial

Implications may be:

- Difficulty in retaining current insurance coverage
- Increase in insurance premiums
- Increase in fixed operating costs
- Reduction in revenue
- Reduction in profits

6.2.9 Legal

Implications may be:

- Possible very high litigation costs.

6.2.10 The Basic CRM Principle

Probably the most important principle and lesson learnt from 50 years of CRM evolution is that an organisation must promote a culture where everyone feels empowered to speak up when they can or whistleblowing, it depends on identifying what the core issue is and resolving it.

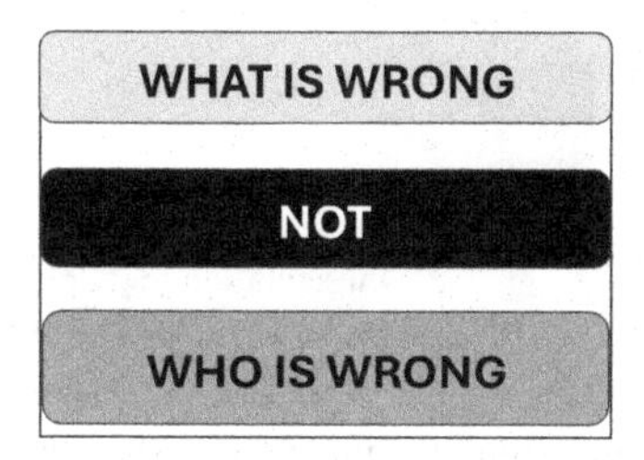

FIGURE 6.1 Basic principle of Crew Resource Management.

6.2.11 Global Rate of Accidents

The graphic in Figure 6.2 illustrates the global rate of aviation accidents involving passenger fatalities per 100 million miles of scheduled commercial air transport operations. It excludes acts of unlawful interference. In the period of time following World War II, aviation technology was evolving at a considerable pace, with the introduction of gas turbine engines, jet engines, and radar as a few examples. There were many accidents thought to be mainly caused by technical failures. However, as the reliability of the technology increased, the accident rate did not improve as expected. Research was conducted on how the crew performed together as a team, and it was found that there was often a mismatch in crew behaviours termed as "Pilot Error". This led to the study of Crew Resource Management (CRM) and Human Factors (HF).

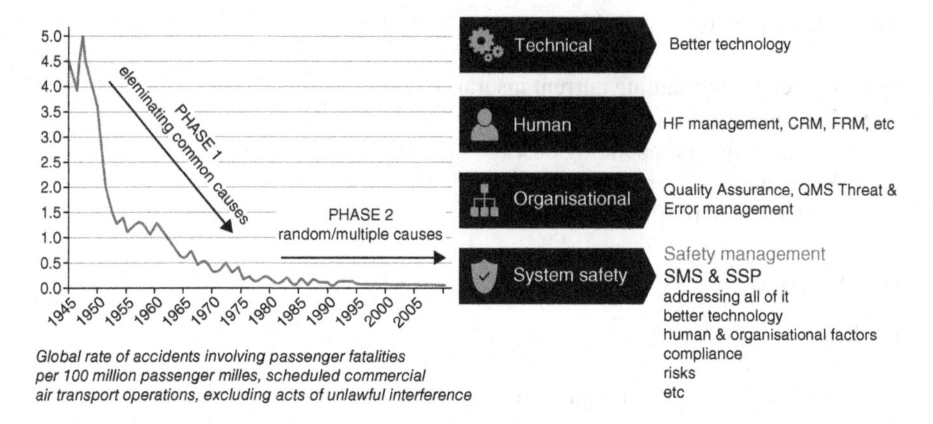

FIGURE 6.2 Evolution of safety thinking related to commercial airline statistics measured between 1945 and 2004.

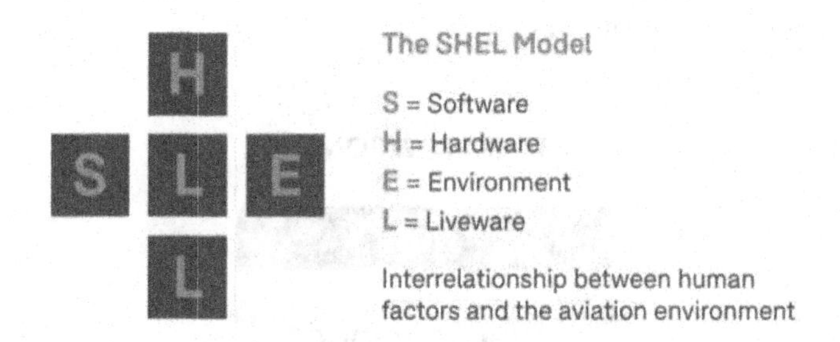

FIGURE 6.3 SHEL model indicating the interrelationship between Human Factors and the aviation environment.

6.2.12 The SHEL Model

The image given in Figure 6.3 introduces the concept of the SHEL model. This indicates the interrelationship between the following:

- Human Factors and the aviation environment
- Software: The rules, procedures, and written documents that form Standard Operating Procedures (SOPs)
- Hardware controls and surfaces, configuration, displays, functional systems, and components.

- Environment: The situation in which the software, hardware, and liveware systems must function the social and economic climate along with the natural environment
- Liveware: The human element flight crews, engineers, maintenance personnel, management, administrative personnel, and other nominated roles such as the Remote Pilot, Accountable Manager, and UAS Operations Manager

6.2.13 IMSAFE

Before operating any aircraft or UAS, it is essential to know whether you are fit to operate the equipment safely. This is for your own benefit, legal protection, and protecting others. The IMSAFE checklist is extensively used by many industries to conduct a preliminary self-assessment.

6.2.14 Illness

- Are you suffering from any illnesses that may affect your ability to take part in the operation?

6.2.15 Medication

- Are you taking prescribed, or non-prescribed medication that may affect your ability to safely take part in the operation?

6.2.16 Stress

- Are your stress levels at a stage where you would not safely be able to conduct the proposed operations?

6.2.17 Alcohol and Drugs

- Are you free from alcohol or drugs and their effects? This includes controlled substances. The prescribed limit of alcohol (Railway and Transport Safety Act 2003 Chapter 20, part 5—Aviation; Alcohol and Drugs) is 9 micrograms of alcohol in 100 millilitres in the case of breath and 20 milligrams of alcohol in 100 millilitres in the case of blood.

6.2.18 Fatigue

- Are you well rested?

6.2.19 Eating

- Are you fed, watered, and ready to go?
- Have you got sufficient food and water to last the duration of the operation?

FIGURE 6.4 IMSAFE Fitness to Fly checklist.

6.2.20 CRM Training

The following list represents an example of the latest subjects that require training over a four-year rolling cycle following the TRM principle.

NOTE: In commercial aviation, legislation requires a three-year rolling cycle; however, the syllabus and content are increasing each year, making it impractical to cover all subject matter in sufficient depth to provide a quality training.

Generic

- Integrated Safety Management Systems—covered in Part One
- Information Acquisition and Processing
- Leadership and Management
- Personality, Cultural, and Generational Differences
- Skill, Error, and Reliability
- Human Performance and Limitations
- Workload Management
- Decision-Making.
- Wellness, Stress, and Stress Management
- Sleep and Fatigue

Job-specific

- Situational Awareness
- Threat Error Management
- Monitoring and Intervention
- Automation Philosophy and Use
- Surprise and Startle Effect
- Resilience Development
- Assessment and Behavioural Markers
- Accident/Incident Investigation

6.3 GENERIC HUMAN FACTORS

My treatment of this subject is based upon many years of practically teaching and applying Crew Resource Management/Team Resource Management/Human Factors principles in the aircraft, flight simulator, and the classroom. Most of the opinions I express are based upon the feedback and reactions I have noted from students. There is still much scepticism and reluctance for those working in this industry to acknowledge many of the findings and who still consider the subject as more of an academic regulatory box-ticking exercise than the real "missing link in aviation safety". However, I am pleased to say that in recent years, attitudes have begun to change rapidly, and I no longer feel that I am beating my head against the wall. The challenge is to strike a balance between the academic and the practical while still making the subject interesting and enjoyable. Over this period, my main conclusion has been the need to offer training in this subject to *all* employees in an organisation irrespective of their specific discipline which I classify as Generic Human Factors. This should be undertaken with mixed-attendance members from all departments including senior management, focussing upon those Human Factors we all experience. This should then be supplemented with the job-specific Human Factors which form the regulatory requirement for Pilots and engineers. The concept is described as TRM (Team Resource Management) and where operators have embraced this concept, the results have been exciting and impressive. The topics contained in this generic section represent some of the common factors which influence human behaviour irrespective of your occupation. The concept of Team Resource Management (TRM) is an evolution of Crew Resource Management (CRM) which evolved predominately for Pilots and engineers. It is designed to integrate *all* organisation members to discuss, share, and understand their own behaviour and how these can influence group behaviour. The theory is that this mutual understanding fosters a better team spirit with all the associated advantages of better safety culture and company efficiency. I have arranged the *generic* subjects in what I believe is a logical progression suggesting we *all* have a brain, albeit wired slightly differently, but maybe do not appreciate how to look after it and use it to its maximum potential. It is often interesting to understand why each of us has a different view of the world and how that impacts upon our values and behaviour. The main take away is to try and understand and respect why each of us behaves as we do, and how that influences our individual and behaviour within specific groups.

7 The Basic Cognitive Functions

7.1 INFORMATION ACQUISITION AND PROCESSING

7.1.1 OVERVIEW

In this chapter, we investigate a very brief and basic introduction to the way we are wired (all slightly differently) and consider the differences between perception and reality. It is so easy to believe that everyone else has the same sense of the World and what we think we know and what we really know. Having even just a simple understanding of how we gather, process, and use data/information/sensation allows us to rationalise how our individual and group behaviours may be formed and explain our reaction to those sensation stimuli.

Have you ever stopped to just think how fantastic the Human Brain really is? Maybe to keep healthy you watch your diet, exercise regularly to maintain your physical health, and as a valuable side effect assist with your mental health. But do you ever allocate sufficient time to meditate, just think about thinking, or just wonder how much available capacity your brain possesses and indeed how much of that potential you actually use.

This chapter attempts to pose some self-awareness/assessment questions to encourage you to consider what we actually think about and some of the mechanisms impacting upon how much we really understand.

We will consider the differences between sensation, perception, and reality. What is real? A recent article in *Scientific American* suggested that the memory capacity of the Human Brain was estimated as 2.5 petabytes. A petabyte is 1,024 terabytes or 1 million gigabytes. According to one study, 1 petabyte is a much data as the whole internet. We now believe that the brain can accommodate more than ten times more data than previously believed (www.medanta.org).

The brain receives signals from all over the body, and these are transmitted by neurons or nerve cells throughout the central nervous system. The health of these networks between neurons and synapses determines our memory capacity. Each neuron can link up to 10,000 others.

Our memory comprises:

- **Sensory Memory**—remembers sensory information after the stimulus ends.
- **Short-Term Memory**—recalls short-term information for about 30 seconds.

DOI: 10.1201/9781032620220-9

- **Long-Term Memory**—stores the majority of our experiences and is thought to be limitless.

The signals the body receives through the senses are known as sensation. Traditionally, we think of only five senses:

- Sight
- Hearing
- Touch
- Taste
- Smell

However, some scientists suggest that we could possess additional senses which we no longer use and lie dormant.

After birth as the brain develops rapidly, these senses (sensations) are given context by our surroundings, the people we interact with, and our ability to learn. The Human Brain is thought to be unique on our planet due to its ability to synthesise various data in ways that allow original thoughts and concepts to develop.

These environmental experiences largely regulate and modify our perceptions. Consider that your perception of any experienced event could be very different to that of someone else. If you accept that argument, then it follows that we may all have slightly different opinions of what is reality!

7.1.2 So How Does It Work?

When we perform any task, stimuli enter our brain through the five senses, and the brain chooses what it needs to complete a task and adds to the working memory. Any information not related to the task is discarded. **Working Memory** and **Long-Term Memory** work together to allow us to perceive the World around us. A combination of memory and perception allows you to make a decision. Your degree of alertness and focus (attentiveness) you have determines how much information you are able to access which is proportional to the quality of your decision-making. By executing your decision, your brain receives feedback in the form of further stimuli, and the process begins all over again.

Sensation and perception are not the same thing. The brain receives raw perceptual data and then processes it using perceptual skills and memory. Completing a clear picture from incomplete information is known as perception. We have to learn to perceive, and this may explain why different people are able to have a different perception of the same sensation. This is a fundamental factor in the understanding of Human Factors, which should encourage everyone to often question their perception of events with the understanding that they could be mistaken! What we actually perceive is often dependent upon the context with which the perception is experienced. This is known as Contextual Perception.

At the end of this chapter, there are links to several video presentations which deal with illusions, and how the brain can be tricked into making erroneous perceptions and assuming they are in fact reality.

7.1.3 Memory

There are three main components of human memory:

- The Sensory Register
- Working Memory
- Long-Term Memory

If we think of our memory as having slots or bins to store information usually, we have about seven slots available, but, depending upon our levels of alertness and wellness, this can range from five to nine slots although specialist Memory Masters can train their brain to contain many more as well as how much information can be placed into each slot. A good analogy is to consider how you store information on your computer. After compiling several documents on the same subject, you may decide to tidy up your desktop and place several files into a folder. It is then easier to make a logical search as to which folder may contain a specific file you may require. Another concept may be to think of how you find a particular book title in a library. First you locate an appropriate subject matter section and then search alphabetically or numerically for the desired publication. It is much easier to locate a folder which contains several files than go directly to that file. Depending upon training or experience, different persons will maybe file information in a different sequence into different memory slots which may explain why their ability to retrieve information of high complexity at an amazing speed appears so easy for them. The efficiency of your storage system will free up more slots to accommodate data pertaining to more tasks. Everyone is limited by the same capacity of **Working Memory** but, with practice, can use this data more efficiently.

If all of your working memory slots are full and you input an additional volume of data, something will fall out. Working Memory is also time limited—normally to about 30 seconds. If you are not consciously focussing on a piece of information after that 30 seconds it will be lost, and you will have to make a conscious effort to repeat that same thought.

Long-Term Memory is an infinite store of facts, procedures, and all your lifetime experiences. Research suggests that we never forget what is stored in our Long-Term Memory; however, we may have problems retrieving the data unless we frequently practise accessing and using this information. The "Tip of the Tongue" phenomena is often when we cannot access a memory straight away and have to trawl through our long-term folders to find the correct pathway to a specific file.

7.1.4 Filling in the Blanks

One interesting fact to consider when thinking about information processing is the phenomenon that humans frequently display whereby if they are trying with difficulty to recall some information completely and accurately, they fill in the blanks with a perception of what they think they perceived. This is dealt with again in the Chapter 8. There is a medical condition rarely understood but experienced by some partially sighted persons and known as **Charles Bonnet Syndrome (CBS).**

7.1.5 Charles Bonnet Syndrome (CBS)

When we investigate the sensation of sight, many presume that see through our eyes. In fact, what actually happens is that light signals are reflected onto our retinal cells in the eye and transferred by the optic nerve to the brain where an image is sensed. This image actually enters inverted, and the brain converts the signal into the correct orientation that our brain actually processes leading to us gaining a perception of what we are looking at.

Sadly, for those partially sighted with CBS, they have a degraded set of retinal cells causing a vast reduction in data transferred to the brain and consequently a greatly reduced perception of what they are looking at. In certain light conditions, this causes the brain to fill in the blanks often with strange memory data not related to the actual reflected data they are unable to process. This often frighteningly causes them to perceive processed vision of imaginary phenomenon from deep in their memory as opposed to what is actually before them. It was long thought that these hallucinations were caused by mental health problems, but that is not the case. Some with acute CBS think they are seeing dead people, dragons, and monsters whereas others see miniature people or children and have learnt to make friends with their visions. These hallucinations are related quite often to stress as well as poor visual acuity. They can be occasional or even frequent and often disturbing, but the root cause is due to the brain filling in the blanks with an incorrect perception of the corrupted sensation data.

The links contained in Chapter 9 are interesting and represent various aspects which help to explain why we think and behave the way we do. I often show these during classroom training, and every time I see them they still surprise me!

7.2 DECISION-MAKING

7.2.1 Overview

We make decisions constantly but maybe never consider how we perform these functions. By deconstructing a few of the basic methods, it may help us to understand how we could improve the quality of this process by considering some common biases and factors which may negatively affect our abilities and outcomes resulting from such decisions.

We are constantly making decisions about all sorts of matters, but have you ever considered the different types of decisions and how you could improve the quality and outcome of the choices you make?

In this chapter, we will analyse some of the contributory factors which impact upon our decision-making ability and performance.

Our brain tries to make each thought process as short as possible if it can and therefore use less attention (mental effort).

We can illustrate the basic types of decision-making using the continuum diagram given next:

Analytical decisions	**Quicker decisions**	**Intuitive decisions**
(effortful, long)	**(decreased time and mental effort)**	**(effortless, quick)**

Time reduces towards the right of diagram.

(Mainly conscious processing) **(does it feel right?)**

(mix of conscious and unconscious)

7.2.2 Analytical Decision-Making

Analytical decision-making is used for complex decisions when time is available. Use of decision-making aids such as checklists and following Standard Operating Procedures can simplify the process.

Decision Making is a conscious process requiring:

- Conscious process
- Objective evaluation
- Best chance of optimal outcome
- Less errors
- Cognitively challenging
- Time consuming
- Increased workload
- Requires attention

7.2.3 Quicker Decision-Making

Using shortcuts to come to conclusions relies on the assessment of incoming information (Situational Awareness) and the processing of information. Mechanisms to facilitate this process include:

- Recency of similar decision
- Neglecting underlying information
- Availability of information
- Using small samples of data to extrapolate a decision
- Anchors in the memory from previous experience
- Confirmation bias
- Value or utility of decision

7.2.4 Very Fast Decision-Making

Recognition-primed decision-making (RPD, Klein, 1998) suggests that typical situations are recognised from previous experience. Implications may include:

- Quickly simulating a rapid response if it feels familiar
- Quickly discarding the above if not suitable response
- Speed of decision-making, not necessarily quality
- Not always the optimum decision
- Serial processing of options

We make our decisions based upon a limited subset of all the information potentially available. There are always things that we think are not relevant at the time, but in hindsight, sometimes they are. In high-pressure situations with very strict time constraints, it is necessary to reduce to pressure, workload by make decisions that work, continually monitoring and reviewing the outcome to decide if a different approach is more suitable. Your ability to make good decisions can be affected by many factors competing for your attention or modifying your performance such as:

- Tiredness
- Fatigue
- Distractions
- Dealing with incomplete information and filling in the gaps with non-relevant information.

In order to try to relieve some of the pressure when making decisions, our brains naturally try to simplify the problem using "rules of thumb", previous experience, and other tools which lead to risks in decision-making described as Biases.

7.2.5 Some Common Biases

Two of the most common biases are listed below:

- Automation—the tendency to quickly trust or favour automated decision-making systems even to the extent of ignoring contradictory but correct information.
- Association—linking together facts either inappropriately or incorrectly. We always want to link together new information with what we already know.

8 The Social Skills

8.1 COMMUNICATION

8.1.1 Overview

Communication is the most basic competency needed by everyone to be able to work and live together in harmony with a clear understanding of how to accurately share feelings, knowledge, and simple and abstract concepts. Good communication skills are however not always evident, and many of us wrongly believe we are better communicators than we are in practice. This chapter attempts to give a brief outline showing how we can improve our skills with practice and a basic understanding of the subject.

Being able to communicate effectively is the most important of all life skills. Yet, many of us do not realise how ineffective we are. This chapter aims to deconstruct some of the basic principles of communication that you may have overlooked or taken for granted. As you read the suggestions in this chapter, you may think this is just simplistic and obvious common sense—because it is! Sadly, common sense is often not so common. And we often overcomplicate matters in the misguided belief that we are more sophisticated or need to present a facade of being more complex than we actually are. Communications skills can be easily practised and improved with only a little effort. Active listening plays a massive part in this process and is usually the area where we are deficient.

Oxford Dictionary definition:

> "The imparting or exchanging of information by speaking, writing or using some other medium".

Latin root: communicatio(n) from the verb communicare—to share.

It has been suggested by www.skilsyouneed.com that time spent communicating may be classified in percentage terms as:

70%—communicating
30%—not communicating

This was further estimated in 2012 as being subdivided into:

Listening—45%
Speaking—30%
Reading—16%
Writing—9%

DOI: 10.1201/9781032620220-10

With the increase in use of social media, these figures may have changed!

The Communications Loop may be formed as given next:

What do we need to deliver a message?	A Sender
What do we have to do to that message?	It needs to be encoded, i.e. converted into a suitable medium to allow for transmission.
Where does that encoded message go now?	A communications channel or data pathway.
What is that suitable medium?	For example: Telephone/video/computer

The message must now be decoded.

So that it can be received by a Receiver.

Once the message has been sent, how do we know we have received it correctly?

There must be an acknowledgement or *feedback* to close the loop.

Although you may think this loop description is obvious, the analysis of incorrectly understood information is usually the result of a breakdown in the continuity of this loop. Breaking the loop at any point can corrupt the integrity of any message. The most frequent reason for a misunderstood transfer of any information results from a lack of objective feedback.

Consider examples of misunderstood messaging you may have experienced and then try to analyse where in the Communications Loop the breakage occurred.

8.1.1.1 Breaking the Loop

What can happen to our ability to clearly understand a message if we break the Communication Loop at any point?

We lose the entire content of the message or only understand part of it.

What do all humans do if they only hear or understand extracts from a complete message?

They fill in the blanks with their own imagination or a perception of what they think they heard!

How can we improve our listening skills?

As suggested in the previous section, most of us do not really listen attentively enough. We often rush to say what we want to say without listening or considering what we have just had explained to us.

Good principles of listening are as follows:

- Stop talking
- Prepare yourself to listen
- Put the speaker at ease
- Remove distractions
- Empathise
- Be patient
- Avoid personal prejudice
- Listen to the tone

- Listen for ideas
- Wait and watch for no verbal communication

What are some of the barriers to good communication?

- Poor environment
- No interest
- Distractions
- Trying to multitask
- Personal state/physical and mental health
- Preconceived ideas or biases
- Closed mind
- Prejudices

Some other barriers are as follows.

Personal factors:

- Does the Sender or Receiver communicate impassively?
- Is it possible to detect any attitudes or aspects of personality which may influence perception?

What style of communication is being used?

- Aggressive
- Passive
- Friendly
- Angry
- Dismissive
- Conciliatory
- Informative.

Perceptual factors are as follows:

- Method/motivation of transmitter or receiver: Is there a clear or hidden reason for the message being delivered?
- Interest of transmitter/receiver: How interested are the parties in being involved in this process? Do they wish to be involved or are they being pressured into it?

Conceptual factors are as follows:

- Understanding: Do all parties in the exchange of information understand the subject matter being discussed from a common level of concept?
- How are any nuances relating to the style of delivery perceived to affect the understanding?
- Knowledge Gradient: Is one of the parties coaching the other?

Practical factors are as follows:

Environment: Is the environment conducive to a clear exchange of information?

- Is it noisy?
- Is it comfortable?
- Is there enough privacy?
- Is it safe?

Structure: Do the concepts being expressed follow a clear standardised structure or are they being randomly expressed?

Message delivery time: Are messages being exchanged within a suitable time frame, allowing all parties sufficient time to assimilate the information?

Feedback: Is there sufficient feedback between parties to allow everyone to confirm that they all understand what is/has been discussed?

Additional influential factors are as follows:

Cultural: The ideas, culture, and social behaviour of a particular group of people or society.

Accent: Does the sender or receiver have a strong foreign or regional accent?

Dialect: Does the sender or receiver have a strong regional dialect?

Mannerisms: Does the sender or receiver have any distracting mannerisms?

Body Language: Does the sender or receiver display any specific attitudes through their body language?

8.1.1.2 Review

After a lifetime of meeting, managing, training, and befriending people from extremely diverse cultures, religions, social and economical classes, I have learnt that the simple concepts of communication combined with an open minded, friendly disposition showing equal respect and understanding for the differences and similarities different people possess, it is quite easy to go anywhere in the world and communicate effectively. Just watch anything with Michael Palin who is one of the most extraordinary communicators, and you will recognise these basic principles expertly utilised.

For successful communication, you MUST HAVE:

- **A SENDER**
- **A RECEIVER**
- **A COMMON LANGUAGE**
- **THE SENDER MUST CLEARLY STATE THEIR MESSAGE**
- **THE RECEIVER MUST LISTEN CAREFULLY AND BE ABLE TO UNDERSTAND THE MESSAGE**
- **THE COMMUNICATION LOOP MUST BE CONNECTED AT ALL STAGES**
- **THERE MUST BE FEEDBACK TO CONFIRM THE MESSAGE WAS PASSED SUCCESSFULLY**

8.2 PERSONALITY, CULTURAL, AND GENERATIONAL DIFFERENCES

8.2.1 Overview

As all of us travel more and work in jobs around the World, we interact with more and more people with different backgrounds of experience, religion, culture, values, and aspirations. Developing an understanding of these differences is essential to be able to work together particularly in safety critical industries. With modern society depending on and living with social media and rapidly advancing technology, it has now become obvious that generational differences can be equally important. In this chapter, we will analyse some of these differences.

This subject is one of the more recent that those working in Aviation Industry have been mandated to study. If you analyse some possible reasons:

- Many organisations subcontract their operations to other diverse organisations around the globe
- Living in a multi-cultural world as well as having many benefits can also create many issues
- Diversity itself results from different cultural norms, religious beliefs, customs, ethnic origins
- Advances in technology are creating many generational differences due to respective acceptances of computer literacy, reliance upon social media, advances in Artificial Intelligence, and the immediacy of 24-hr non-stop global news availability.

All humans are wired differently and evolve in slightly different ways with different views on life and basic principles of what behaviour is or is not acceptable in their own culture. It is now more important than ever before in our history that if we are to work and live together in harmony, we need to understand, respect, and cultivate diversity. This chapter aims to deconstruct many of those differences. It is based upon scientific research and does not exploit any prejudice or stereotypical ideologies. The fact is that we are all different and exhibit many differences of diversity.

8.2.2 What Makes Us All Different?

To answer this, here are some questions we should ask ourselves.

- What are personality traits?
- How can we identify our personality traits?
- How many different cultures have you experienced?
- What are some obvious cultural differences?
- What are some obvious generational differences?

- How do the two questions above affect our working relationships?
- Are any differences real or perceived?

8.3 WHO DO YOU THINK YOU ARE?

8.3.1 Personality

- Everyone is different. Why?
- Is personality inherited from the culture that the person developed in?
- Where do individual personality traits develop?
- Is the difference between culture and personality to appreciate that personality differences are the characteristics of an individual within their own national group?
- The idea of personality is familiar to everyone, but what does it mean?—Broadly speaking, it describes the patterns of behaviour, thought, and emotion that make a person unique.

8.3.2 Have You Heard of the BIG 5 in Terms of Personality Traits?

Remember the mnemonic **OCEAN.**

Openness—the extent to which you are receptive to novel ideas, creative experiences, and different values.
Conscientiousness—the extent to which you are organised, strategic, and forward planning.
Extraversion—the extent to which you are inclined to experience positive emotions and how attracted you are to social, stimulating experiences.
Agreeableness—the extent to which you are concerned about the feelings of others, and how easily you bond with people.
Neuroticism—the extent to which you react to perceived threats and stressful situations.

8.3.3 Culture

What is it? The way people behave, think, interact towards with each other generally, as well as what motivates them, will partly be a factor of their national and cultural background.

What influences those differences?

- Language
- Communication
- Rituals
- Courtesies
- Roles and customs
- Relationships
- Practices

- Expected behaviours
- Values
- Thoughts
- Manners of interacting

8.3.4 Areas of Common Cultural Differences

These may include:

- Perception of time
- Family structure
- Hygiene
- Clothing
- Personal space
- Religious beliefs/practices
- Manners
- Body language
- Gender roles
- Reaction to problems
- Food/mealtimes
- Greetings
- Modes of communication

8.4 CONFLICT RESOLUTION (DEALING WITH DIFFICULT CUSTOMERS AND COLLEAGUES)

8.4.1 Overview

This section has been added because the analysis of so many accidents/incidents and case studies highlight how often lack of a JUST CULTURE, various personality clashes, and conflicts have been identified as the root cause of an occurrence. Thus, it was considered appropriate to include some background information to help readers avoid or resolve different forms of conflict.

8.5 CONFLICT

8.5.1 *Oxford English Dictionary* Definition

Conflict means a serious disagreement or argument, typically a protracted one—a state of mind in which a person experiences a clash of opposing feelings or needs; serious incompatibility between two or more opinions, principles, or interests. Within any company, the potential for conflict can be drastically reduced by establishing and maintaining the principles of a **Just Culture**. Everyone who feels confident enough to be able to express a view which highlights a problem or just stimulates a dialogue with the potential to contribute to the resolution of any issue adds considerable value to the company. A Just Culture contributes to a reduction in tension, which leads to a reduction in conflict and provides a suitable roadmap for transition into a smoother resolution process. A large proportion of conflict is based upon fear. This can be real fear or perceived fear, largely of the unknown. These fears can be largely mitigated by having an open culture where important information is clearly disseminated to all. If everyone clearly understands how a company functions and trusts the structures that are there to protect them, then there should be nothing to fear. If everyone is clearly aware of their duties, responsibilities, liabilities, and accountabilities within a disciplined, transparent culture, again this has the effect of drastically reducing potential conflicts. In situations when a conflict cannot be predicted and a mitigating strategy planned and executed, there are some basic rules which can be practised and adopted as a first line of defence. Some of those techniques are listed in the next sections.

8.5.2 Forcing

Also known as competing. An individual firmly pursues his or her own concerns despite the resistance of the other person. This may involve pushing one viewpoint at the expense of another or maintaining a firm resistance to the other person's actions.

Examples of when this technique may be appropriate are:

- In certain situations when all other less forceful methods do not work or are ineffective
- When you need to stand up for your own rights and resist aggression and pressure
- When a quick resolution is required and using force is justified (e.g. a life-threatening situation to stop aggression)
- As a last resort to resolve a long-lasting conflict

Possible advantages of forcing:

- May provide a quick resolution to conflict
- Increases self-esteem and draws respect when firm resistance or actions were a response to an aggression or hostility

Some possible limitations:

- May negatively affect your relationship with the opponent in the long run
- May cause your opponent to react in the same way, even if the opponent did not intend to be forceful originally
- Cannot take advantage of the strong sides of the other side's position
- Taking this approach may require a lot of energy and be exhausting to some individuals

8.5.3 Win-Win

Also known as problem solving, requires working together to find a common ground.

Examples of when this technique may be appropriate are as follows:

- When consensus and commitment of other parties are important
- In a collaborative environment
- When addressing the interests of multiple stakeholders is required
- When a high level of trust is present
- When a long-term relationship is important
- When you need to work through hard feelings, animosity, etc.
- When you do not want to have full responsibility

Possible advantages of win-win:

- Leads to solving the actual problem
- Leads to a win-win outcome
- Reinforces mutual trust and respect
- Builds a foundation for an effective collaboration in the future
- Shared responsibility of the outcome
- You earn the reputation of a good negotiator
- The outcome can be less stressful

Some possible limitations:

- Requires a commitment from all parties to look for mutually acceptable solutions
- May require more effort and more time
- May not be practical when time is crucial, and a quick resolution is needed
- One or more parties lose their trust in an opponent

8.5.4 Compromising

Here, both parties are looking for mutually acceptable solutions.

Examples of when this technique may be appropriate are:

- When goals are only moderately important
- To reach a temporary settlement on complex issues
- To reach expedient solutions on important issues
- As a first step when the involved parties have not yet developed trust
- When collaboration or forcing do not work

Possible advantages of compromise are as follows:

- Faster issue resolution
- May find a temporary solution
- Lowers the levels of tension and stress resulting from the conflict

Some possible limitations are as follows:

- May result in neither party being satisfied with outcomes
- Does not contribute to building long-term trust
- May require close monitoring and control to ensure that the agreements are met

8.5.5 Withdrawing

Also known as avoiding.

Examples of when this technique may be appropriate are as follows:

- When the issue is trivial and not worth the effort
- When more important issues are pressing
- When it is not the appropriate time or place to discuss the issue
- When you need more time to think about your position
- When you see very little chance of addressing your concerns
- When you have to deal with hostility
- When you are unable to handle the conflict

Possible advantages:

- When the other person is forcing and attempts aggression, it may be better for you to wait and postpone your response to a more appropriate time
- Low-stress approach when conflict is short
- Allows time to focus upon more urgent issues
- Allows time to compose a better argument strategy

Some possible limitations:

- May lead to losing or weakening your position
- May negatively affect your relationship with one of your own team if they expect action from you

8.5.6 Smoothing

Also known as accommodating.

Examples of when this technique may be appropriate are as follows:

- When it is important to provide temporary relief or buy time to compose your argument.
- When the issue is not as important to you as the other person.
- When you accept you are wrong.
- When continued conflict would be detrimental, and you have no choice.

Possible advantages are as follows:

- Sometimes protects more important interests while giving up on less important ones.
- Gives the chance to reassess the situation from a different perspective.

Some possible limitations:

- There is a risk to be abused. Opponent may constantly take advantage of your tendency to be accommodating.
- May negatively affect your confidence in future dealing with aggressive opponents.
- More difficult to transition to a win-win situation in the future.
- Some of your colleagues may dislike your response, and you will lose their respect.

8.5.7 General Principles

What we have seen in the previous strategies is the context of employing different strategies to resolve conflict. However, one of the golden rules is to avoid the conflict in the first place. For example you are a Remote Pilot operating for a client who decides to question your decision to delay operating the mission due to poor weather conditions. Do not rise to the bait. Quietly and politely inform him that you are following company regulations which are not up for discussion. If he still maintains an aggressive manner, again politely and quietly ask him to contact your Company Operations Department and speak to your boss and slowly walk away. This is not your job. Answering back in a defensive manner will only escalate the conflict.

Walk Away.

8.6 LEADERSHIP AND MANAGEMENT

NB: This topic generally belongs in this section dealing with the Social Skills; however, since this publication is dealing principally with specific management skills, we have devoted a distinct Part Five to deal with the subject including a practical desk top exercise.

9 The Practical Application

9.1 HUMAN PERFORMANCE AND LIMITATIONS

9.1.1 Overview

In this chapter, we will analyse some of the factors which impact and affect our ability to perform tasks and suggest some practices which may facilitate the efficiency and safety. Many of these practices are common sense and sometimes so obvious that it is even more surprising that we do not always recognise them. Therefore, it is often a good idea just to review and go back to simple basics to help us understand what affects our performance and limitations. We have already seen the **SHEL** model outlined in this publication—Software, Hardware, Environment, Liveware. In this chapter, we will focus upon:

Environment—the situation in which the Software, Hardware, and Liveware systems must function. The social and economic climate along with the natural environment.

Liveware—The human element: Flight Crews, engineers, maintenance personnel, management, administrative personnel, and other nominated roles such as the Remote Pilot, Accountable Manager, and UAS Operations Manager.

The mnemonic **PEAR** is used to recall four considerations for assessing and mitigating human factors in aviation.

People who do the job.
Environment in which they work.
Actions they perform.
Resources necessary to complete the work.

The following tables illustrates the component factors.

9.1.2 People

Physical factors	Physiological factors	Psychological factors
Physical size	Nutritional factors	Workload
Gender	Health	Experience
Age	Lifestyle	Knowledge
Strength	Fatigue	Training
Sensory limitations	Chemical dependency	Attitude
Endurance	Alcohol usage	Mental or emotional state

DOI: 10.1201/9781032620220-11

9.1.3 Environment

Physical	Organisational
Weather	Personnel
Location inside or outside	Supervision
Workspace	Labour Management relations
Shift	Pressures
Lighting	Crew structure
Sound level	Size of company
Vibration level	Profitability
Safety	Morale
Comfort	Corporate culture

9.1.4 Actions

Steps to perform tasks
Sequence of activity
Number of people involved
Communication requirements
Information control requirements
Knowledge requirements
Skill requirements
Attitude/behavioural requirements
Certification requirements
Inspection requirements

9.1.5 Resources

Engineers	Pilots
Procedures, work cards	Standard operating procedures
Technical manuals	Technical manuals
Other people	Other people
Test equipment	Aircraft
Tools	Clothing
Computers' software	Computers/software
Paperwork, signoffs	Paperwork
Ground handling equipment	ATC
Work stands/lifts	Operations
Fixtures	Crew members
Materials	Commercial pressure
Task lighting	Practice
Training	Training
Quality systems	Quality systems

In the UK, one excellent resource is the Health and Safety Executive. They monitor, regulate, and research into developing and maintaining standards of safety in the workplace.

9.2 GENERIC ERROR MODELLING SYSTEM (GEMS)

This system developed by Professor James Reason proposes that errors may be classified as:

Skill based—slips and lapses usually caused by misplace or in attention to the task.
Rule based (mistakes)—usually resulting from following inappropriate or deficient rules or failure to follow accurately these rules.
Knowledge- based mistakes due to incomplete/inaccurate understanding of system, confirmation bias, overconfidence, cognitive stresses.

Generally, within an organisation operating a Just Culture, it is accepted that because we are fallible humans, errors will occur. So long as these are reported, and the organisation can learn from these occurrences, they should not cause any punishment as opposed to deliberate violations of the rules or acts of sabotage which are not acceptable and will be punished. These principles form the basis of most contracts, terms and conditions, and disciplinary procedures.

9.3 WORKLOAD MANAGEMENT

9.3.1 Overview

In this section, we will consider the impact of workload management and its affect upon safety and efficiency. Again, the principles are common sense and quite obvious, but it sometimes is necessary to step back and revisit why pacing oneself is a significant factor in the workplace to reduce stress, focus upon the essential tasks, and maybe stop to think before acting in haste.

In all the years that I operated as a Pilot, trainer on the aircraft, and in-the-flight simulator, the most common issue I observed with colleagues who had some issues in the work environment would be poor time/workload management.

When you are stressed, one of the first things to cause problems is usually losing an accurate perception of time. The old adage used in training is when you experience an emergency, just sit on your hands, take some deep breaths, and count to ten. Nothing could be truer. Time management includes a propensity for many aircrew to try and perform too many tasks without sensible delegation and then cram the workload into an inadequate time frame.

Factors with a major influence causing high workload include:

- How difficult or demanding the task is
- The number of concurrently running tasks being attempted
- The amount of switching between one type of task and a different task
- The time available to perform the task or the speed of action required

The concept of multitasking is actually a myth. Humans can only really focus upon one task at a time to perform it with maximum efficiency.

The best method to efficiently manage time is to develop your team work to the extent that you can trust and rely on your colleagues to share the workload and feel confident to delegate and not always micromanaging. This can be best developed by having standardised structures and operational procedures where investment in training is the key to improving workload management and ensuring operating with a Just Culture where everyone understands their duties and carries them out with a calm and confident professionalism.

Specific factors to improve workload management are the basic competencies of knowledge, skill, and behaviour/attitude. Obviously, efficiency is directly proportional to the reduction of routine and unexpected errors. An interesting study described below investigates the relationship between errors and how they may become escalated under certain conditions.

9.4 RISK FACTORS IN THE TECHNIQUE FOR HUMAN ERROR RATE PREDICTION (THERP) (SWAIN AND GUTTMANN 1993)

After a major meltdown of an American nuclear facility on a three-mile island in Pennsylvania in 1979 of proportions and release of radioactive pollution similar to that experienced in Chernobyl, the investigation arrived at several conclusions.

The results have since been shown to be equally relevant with identical impacts upon many safety critical operations including aviation.

Condition	Escalation Risk Factor
Unfamiliarity with task	**17 times**
Time shortage	**11 times**
Poor signal/noise ratio	**10 times**
Poor human/system interface	**08 times**
Design–user mismatch	**08 times**

FIGURE 9.1 THERP—Technique for Human Error Rate Prediction.

Condition	Escalation Risk Factor
Irreversibility of errors	**08 times**
Information overload	**06 times**
Negative transfer between tasks	**05 times**
Misperception of risk	**04 times**
Poor feedback from system	**04 times**
Inexperience (NOT lack of training)	**03 times**
Educational mismatch of person with task	**02 times**
Disturbed sleep patterns	**1.6 times**
Hostile environment	**1.2 times**
Monotony or boredom	**1.1 times**

FIGURE 9.1 (Continued)

It is interesting to note that the first two observed conditions in the table increase the risk of error by 17 and 11 times, respectively.

9.5 RESILIENCE DEVELOPMENT

9.5.1 Overview

Resilience development is an interesting subject which can help explain and develop some techniques to mitigate some of the more extreme reactions to the often-experienced stimuli found in everyday life. In a safety critical operation, a combination of good Situational Awareness often derived from an understanding of Threat Error Management can help mitigate such reactions as described in Surprise and Startle. As with all these processes, understanding can improve our ability to recognise and react in a more controlled manner.

Resilience is the ability to return to the original position/form after being altered in some way. It also refers to the ability to recover from factors such as illness, anxiety, and depression. With good training, employees can be taught techniques and strategies to improve their resilience with regard to factors explained in next sections.

9.6 MENTAL FLEXIBILITY

- Having mental flexibility—necessary to recognise critical changes
- Evaluate and review their judgements and modify their behaviour to unique situations
- Reduce always relying on standard solutions and fixed prejudices
- Having the ability to modify preconceptions, assumptions, and perceptions

9.7 PERFORMANCE ADAPTION

- Adjust to prevailing situations/conditions.
- Mitigate frozen behaviours.
- Mitigate overreaction.
- Prevent inappropriate hesitation.

Building resilience is a personal experience as we are all wired differently and require varying strategies to suit each individual. Many of these differences are related to cultural factors as described in Section **Personality, Cultural and Generational Differences.**

Some suggested methods to build resilience are as follows:

- Make connections—good relationships.
- Avoid seeing crises as problems with no solution.
- Accept and embrace change as a part of life.
- Set achievable goals and progress them diligently.
- Be decisive when acting on adverse situations.

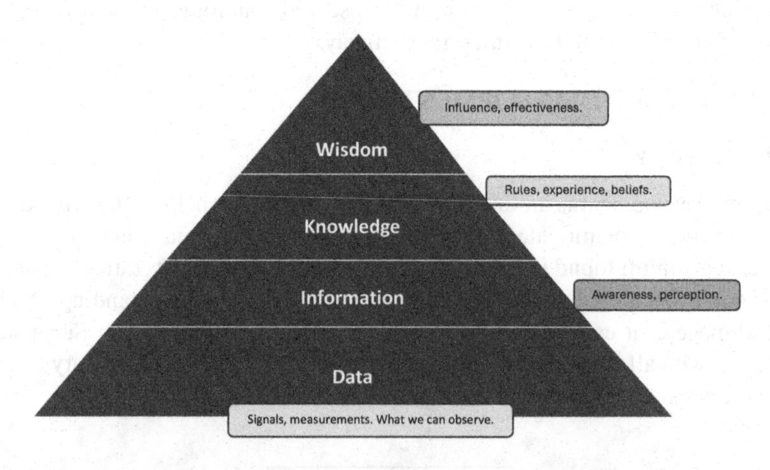

FIGURE 9.2 The DIKW (Data, Information, Knowledge, Wisdom) pyramid.

- Embrace personal development.
- Cultivate a positive self-view.
- Keep events in perspective.
- Develop positive and optimistic outlooks on life.
- Pay attention to your own needs—look after yourself.

9.8 GOOD EXPERIENCE AS A REINFORCEMENT FOR PERFORMANCE ADAPTION

In a way similar to what we will see in **Section Situational Awareness,** performance adaption can be improved with good training and building up a reserve of positive subject knowledge, skill, and experience balanced with positive attitudes and behaviours.

10 The Wellness Skills

10.1 STRESS AND STRESS MANAGEMENT

10.1.1 Overview

Stress and stress management are now considered to be one of the most significant factors in days lost in the workplace. It is a fact of life that all organisations and employees need to be aware of and take appropriate mitigating steps to avoid as well as exercising suitable processes to prevent.

Working days lost in 2022/23 were estimated as 35.2 million (3.7 million owing to non-fatal workplace injuries, 31.5 million owing to workplace ill health)

Stress, depression, or anxiety and musculoskeletal disorders accounted for the majority of days lost due to work-related ill health in 2022/23, giving rise to numbers 17.1 million and 6.6 million, respectively. On an average, each person suffering took around 15.8 days off work. This varies as follows:

- 6.6 days for injuries
- 17.8 days for ill health cases
- 19.6 days for stress, depression, or anxiety
- 13.9 days for musculoskeletal disorders

Stress is a fact of life and affects all of us to varying degrees. It is caused by our reaction to various conditions known as stressors. It is important to live with an optimum degree of stress to allow us to function efficiently. There are two main classifications of stress:

Eustress—Considered as "Good Stress", allowing us to feel energised by being adapted by the body and stimulating us to become energised.

Distress—Considered as "Bad Stress", it is indicated when we feel out of control and unable to cope. Stress develops when perceived ability to perform a given task does not match the demands. It can be physiological (physical) and psychological (mental) response impacting the individual's performance.

10.2 THE STRESS BUCKET

A good analogy to describe the effects of stress is to consider a bucket filled to different levels with stressors. We are all different and have different capacity buckets and varying baseline contents/levels of stress.

DOI: 10.1201/9781032620220-12

The level is constantly changing, and at any one time, it is not always possible to know where your actual level is. There is no content gauge to help us. So long as we manage to cope with events, the bucket will not overflow. Without this knowledge, it is helpful to consider that the bucket is fitted with a tap to reduce the levels at an appropriate time and rate. This tap may be regulated by adopting routine and regular lifestyle strategies to prevent overflow.

Some suggested mitigating strategies include:

- Exercise
- Good healthy diet
- Activities such as playing a musical instrument or doing something different outside your comfort zone
- Writing a list of your problems
- Sharing your problems with someone else

The different levels of severity of the stress are classified as:

- Normal
- Acute—sudden and unexpected pressures such as forgetting where you put your car keys
- Chronic—A prolonged build-up of specific or cumulative problems such as financial or personal relationship issues

The causes of stress are classified as shown in the next sections.

10.3 ENVIRONMENTAL (PHYSICAL)

- Noise
- Vibration
- Temperature
- Humidity
- Glare
- Life (**Psychological**)—emotional, domestic, social, financial.
- Life (**Physiological**)—hunger, thirst, pain, lack of sleep, fatigue.
- Reactive—physical and mental reaction to a significant event.
- Car accident/incident

NB: Refer to the "Lifestyle Stress Calculator" table contained later in this chapter.

10.4 ORGANISATIONAL

- Workload and autonomy
- Relationships with others
- Lack of career development
- Pay inequality
- Bureaucratic processes

10.5 SOME FURTHER EFFECTS OF STRESS

- **Omission**
- **Error**
- **Queuing**—Sequentially delaying necessary actions in an inappropriate
- order of attention priority.
- **Filtering**—Rejection of particular tasks because of overload.
- **Approximation**—Making approximations in technique to cope with all the tasks required in a short-term interval.
- **Coning of attention**—When stress increases, attention scan closes into a smaller field of awareness.
- **Regression**—Under stress, behaviour may regress to the earliest learnt.
- **Escape**—The ultimate response to extreme levels of stress is to give up or freeze.

10.6 FIFTY COMMON SIGNS AND SYMPTOMS OF STRESS

1. Frequent headaches, jaw clenching, or pain
2. Gritting, grinding teeth
3. Stuttering or stammering
4. Tremors, trembling of lips, hands
5. Neck ache, back pain, muscle spasms
6. Light headedness, faintness, dizziness
7. Ringing, buzzing, or "popping sounds"
8. Frequent blushing, sweating
9. Cold or sweaty hands, feet
10. Dry mouth, problems in swallowing
11. Frequent colds, infections, herpes sores
12. Rashes, itching, hives, "goose bumps"
13. Unexplained or frequent "allergy" attacks
14. Heartburn, stomach pain, nausea
15. Excess belching, flatulence
16. Constipation, diarrhoea, loss of control
17. Difficulty in breathing, frequent sighing
18. Sudden attacks of life-threatening panic
19. Chest pain, palpitations, rapid pulse
20. Frequent urination
21. Diminished sexual desire or performance
22. Excess anxiety, worry, guilt, nervousness
23. Increased anger, frustration, hostility
24. Depression, frequent or wild mood swings
25. Increased or decreased appetite
26. Insomnia, nightmares, disturbing dreams
27. Difficulty concentrating, racing thoughts
28. Trouble learning new information
29. Forgetfulness, disorganisation, confusion

30. Difficulty in making decisions
31. Feeling overloaded or overwhelmed
32. Frequent crying spells or suicidal thoughts
33. Feelings of loneliness or worthlessness
34. Little interest in appearance, punctuality
35. Nervous habits, fidgeting, feet tapping
36. Increased frustration, irritability, edginess
37. Overreaction to petty annoyances
38. Increased number of minor accidents
39. Obsessive or compulsive behaviour
40. Reduced work efficiency or productivity
41. Lies or excuses to cover up poor work
42. Rapid or mumbled speech
43. Excessive defensiveness or suspiciousness
44. Problems in communication, sharing
45. Social withdrawal and isolation
46. Constant tiredness, weakness, fatigue
47. Frequent use of over-the-counter drugs
48. Weight gain or loss without diet
49. Increased smoking and alcohol or drug use
50. Excessive gambling or impulse buying

10.7 STRESS CALCULATOR (RISK ASSESSMENT)

For some time now insurance companies have been using many tools to assess risks that they may consider underwriting. As we noted from the chapters about Safety Management Systems, there was the concept of As Low as Reasonably Practical (ALARP). This has largely been driven as a basic component of almost any type of Risk Assessment.

The analysis of stress and respective major stressors has developed a stress calculator tool as included later in the chapter. Different organisations populate the tool with slightly different criteria and classify the results using slightly different thresholds (or final total score). However, to a certain extent, the score is not as important as the assessment of a numerical quantum of the stress and where the stress levels are quantified as:

- Normal
- Acute
- Chronic

We have already seen that depending upon which level of stress one is experiencing, the effects can impact upon the production of several hormones which affect behaviour. What is less understood is the fact that often a cocktail of hormones can be released causing different reactions in different people. This mechanism can connect the impact of mental stress causing major physical reactions or changes in health and behaviour. One of the less understood symptoms of chronic stress is that many of your private thoughts and behaviours very rapidly become irrational, leading to random and often very dark thoughts and actions.

Stress Calculators (Risk Self-Assessment)

Pre-Flight Check: Am I Fit to Fly?

- Could I pass my medical at this time?
- Do I feel well?
- Is there anything wrong with me at all?
- Have I taken any medication within the last 12 hours?
- How many units of alcohol have I consumed within the last eight hours?
- Am I tired?
- Did I get a good night's sleep last night?
- Am I under undue stress?
- Am I emotional right now?
- Have I eaten a sensible meal in preparation for my duties to follow?
- Am I properly hydrated?
- Do I need to drink more water?
- Do I have the right equipment with me to perform my duties? For example, sunglasses, high viz, ear defenders, correct clothing.

LIFESTYLE STRESS CALCULATOR

Event	Stress score	Enter your score
Death of spouse or child	100	
Divorce	60	
Menopause	60	
Separation from living partner	60	
Jail term or probation	60	
Death of close family member	60	
Serious personal injury or illness	45	
Marriage or establishing new life partnership	45	
Fired from work	45	
Marital or relationship reconciliation	40	
Retirement	40	
Change in health of an immediate family member	35	
Work more than 40 hours per week	35	
Pregnancy or partner becoming pregnant	35	
Sexual difficulties	35	
Gain of new family member	35	
Business or work role change	35	

FIGURE 10.1 A standard insurance industry stress calculator.

Event	Stress score	Enter your score
Change in financial state	30	
Death of a close friend (not a family member)	30	
Change in number of arguments with spouse or long-term partner	25	
Mortgage or loan for major purpose	25	
Foreclosure of mortgage or loan	25	
Sleep less than eight hours per night	25	
Change in responsibilities in work	25	
Trouble with in-laws or children	25	
Outstanding personal achievement	20	
Spouse begins or stops work	20	
Beginning or end of school	20	
Changes in living conditions (visitors in the home/ change of roommates)	20	
Change in personal habits (smoking, diet, drinking, exercise)	20	
Chronic allergies	20	
Trouble with boss	15	
Change in work conditions or hours	15	
Moving to new house	15	
Presently in pre-menstrual period	15	
Change in schools	15	
Change in religious activities	15	
Change in social activities (more or less than before)	10	
Minor financial loan	10	
Change in frequency of family get togethers	10	
Have been or just about to go on holiday	10	
Presently in Christmas season	10	
Minor violation of the law	5	
TOTAL SCORE		

Everyone has different stress-coping mechanisms and levels of acceptability.

An average score of 250 or more often indicates that high levels of stress are present.

Studies have shown that stress levels of 300 or more may affect your immune system and have a significant effect upon your physical health.

FIGURE 10.1 (Continued)

10.8 THE GERMAN WINGS A320 FATAL ACCIDENT

This tragic accident has become a watershed moment in accident investigation but sadly with many of the findings being overlooked.

We have spoken about many factors which are stressors including financial pressure. In my experience, one of the most common but ignored factors in commercial aviation is in fact financial pressure. For many years, the Pilots were mainly recruited on the basis of the fact that they were previously from the military, and what distinguish Pilots from many other professions are the levels of enthusiasm, love for the job, and the motivation to accept almost any obstacle just for the love of flying. However, since after the Second World War, many have entered the profession by the route of self-sponsorship or 100% funded cadet training courses operated by major airlines. Many of those airlines have now changed their processes to bond successful cadets, leaving them with training debts in the region of up to £200,000! This was the case of the Co-Pilot in this accident who was sponsored by Lufthansa. Under German law, it is not allowable for an aviation doctor to disclose certain aspects of the health of their patients. Many aspects of mental health (WELLNESS) being particularly challenging to diagnose.

The Co-Pilot owed many thousands of Euros to Lufthansa in sponsorship funds, and it is suggested that due to other underlying mental health issues in addition to financial stress, he was worried about how he would cope if he lost his job due to his deteriorating mental health condition. One of the rarely mentioned aspects of the accident report findings was the inadequate loss of licence insurance provision afforded to many Pilots today which would help mitigate any debts owing in the event of loss of employment.

It is interesting that in 1972 when I obtained my first commercial job as an air taxi Pilot, my salary was £4,000 per year with Loss of Licence coverage of £150,000. Recently, a fully qualified Captain operating offshore in the North Sea can expect a salary in the order of £120,000 while still only having a Loss of Licence coverage in the order of £300,000.

10.8.1 An Example of the Authors Personal Experience

A major finding in this accident was the suggestion that organisations need to design, develop, and immediately mandate confidential Wellness programmes suitable to council and assist the growing number of employees suffering from work-related stress with the specific aim of mitigation of suicide risks.

Many years ago, after an accident in which I lost all feeling down the left side of my body and could only walk with crutches, I experienced such suicidal feelings.

I was at the top of an escalator in a shopping mall with my wife and two children when a group of about six teenagers ran into my wife nearly sending her falling down the escalator and knocking away my crutches so I fell heavily. No one came to our assistance. By the time I arrived at home, something snapped and I seemed to blow a fuse, beginning to feel that as a once big strong man in control, I now was unable to protect my family. These feelings very quickly pushed me over my already chronic stress bucket level into a place where I felt useless and irrelevant and believing that my family would be better off without me. I actually considered ending my life until my dear wife snapped me back into reality by telling me how much she loved me and needed me irrespective of how disabled I may become. She also informed me in a very matter of fact way that if I did take my own life, it would invalidate all our insurance policies!

Sometimes to solution to stress can be very quick and simple!

10.9 PRACTICAL STRESS REDUCTION TECHNIQUE

10.9.1 Part 1

Stand up with your feet about 50 cm apart and your arms by your sides.

Slowly inhale with a slow deep breath while raising your arms up in front of you until they are raised above your head with palms touching. Hold the breath in while you lower your arms stretching them out on each side.

Now slowly lower your arms back down to your sides while breathing slowly out. Exhale attempting to empty your lungs with stale air until you feel the need to breath in again. Repeat the process for about two minutes or until you feel more relaxed.

10.9.2 Part 2

10.9.3.1 5, 4, 3, 2, 1 Technique

1. Concentrate on five things you can see—notice while naming the five things you can see. Focus upon the size, shape, and colour of things you see while thinking how each item relates to the rest of the environment.
2. Concentrate on four things you can touch—focus on things that feel comforting, such as a part of the body, your wedding ring, your coffee cup, or your pen. Focus upon what you feel when you touch each item such as the texture, soft, rough, flexible, heavy, light.
3. Concentrate on three things you can hear—someone talking, a car horn, music playing. Try to listen to the quieter sounds that tend to be in the background unnoticed. Note which sounds bring you a sense of comfort and calm.
4. Concentrate on two things you can smell—which elicit emotions, memory, and even hunger.
5. Concentrate on one thing you can smell and one thing you can taste—taking a sip of your drink.

10.10 SLEEP AND FATIGUE

10.10.1 Overview

We all need sleep but how much and why? What are the impacts of disturbed sleep? In this section, we will investigate some of those answers and what actions we can take to mitigate the effects of lack of sleep.

10.11 SLEEP

We all need sleep to regulate and restore the balance of parts of our Central Nervous System. In a way similar to a computer transferring Random Access Memory (RAM) (momentary memory) to the hard disk, freeing up this memory for the next period of wakefulness.

Most humans need about eight hours of sleep per night to allow the cycle to perform at an optimum level.

Failure to achieve this regular sleep can cumulatively affect:

- Alertness
- Increased reaction time
- Cognitive performance
- Vigilance
- Memory problems

10.12 FIVE STAGES OF SLEEP

- Stage 1—taking around ten minutes, this is the phase between waking and sleeping
- Stage 2—taking around 15 minutes with sleep becoming deeper
- Stage 3—taking about a further 15 min before transition to Stage 4
- Stage 4—after about 90 minutes to transition into Rapid Eye Movement (REM)

REM—this stage is thought to be occurring when the brain organises memory and becomes stronger

The whole cycle of sleep stages 1 to 4 plus REM takes about 90 minutes and repeats about 5 times during an 8-hour cycle.

Some factors impacting sleepiness:

- Previous sleep and wakefulness
- Circadian rhythms
- Early morning decreased performance
- Age (reduces after age 50)
- Alcohol
- Environmental and work conditions

10.13 FATIGUE

Fatigue occurs when the body experiences cumulative effects of disturbed sleep caused by many factors including:

- Too many early starts
- Night shifts
- Long days
- Regular crossing of time zones

10.14 FATIGUE MANAGEMENT

Can be either short-term acute or long-term chronic leading to impaired performance and a reduction of awareness. Some of the symptoms include:

- Tiredness
- Slow reactions
- Diminished motor skills
- Reduced visual acuity
- Reduction of short-term memory capacity
- Impaired concentration

Effects of Fatigue

- Easy distraction
- Reduction in awareness
- Increased errors
- Abnormal mood swings

10.15 MITIGATING THE EFFECTS OF FATIGUE

- Keep fit
- Eat regular balanced healthy meals
- Avoid excess alcohol
- Control aspects of life such as emotional and psychological
- Try to plan and execute regular sleep patterns

It is essential that organisations understand and consider the short- and long-term effects of fatigue upon safety and retention of employees and plan all work patterns appropriately.

10.16 SURPRISE AND STARTLE EFFECT

10.16.1 Overview

This section gives an introduction to a common phenomenon known as the Surprise and Startle Effect, explaining how we may react to shocks caused by unexpected events and how they may impact upon performance, particularly if operating various items of machinery which could result in a hazardous outcome. It suggests some mitigating recovery techniques.

When working in any Safety Critical industry, there should be some attention and training given to understanding the subject of the Surprise and Startle Effect. This should include:

- Effects and causes
- How to recognise them

- How to manage them by developing appropriate automatic behavioural responses
- Strategies to recognise and regain control of situational awareness

The Startle reflex itself can be explained as the reaction to unexpected events such as sudden noises and abrupt exposure to shocking perceptions. It normally triggers within 100 milliseconds after the stimulus and contains subjective (similar to fear or anger) and physiological dimensions, causing a speeding up of your reactions and increased focus of attention. For the first 3 seconds, the reaction is autonomic but may continue for up to 20 seconds. Real or imaginary perceptions of fear can also trigger or exacerbate such reactions. The body can react by blinking the eyes and sometimes by giving a whole-body jerk.

10.17 FLIGHT OR FIGHT

Perception of imminent serious threats causes the brain to release a cocktail of hormones, specifically, adrenaline, which causes an increase in heart rate. Usually, the body will schedule an increase of glucose in the blood system and focus all mental capacities in an attempt to escape or survive the stimulus.

Although this system worked well over millennia while humans reacted to being threatened by a predator or enemy, on its own, the reaction is not easily adaptable to many of the shocks and stresses of modern living, particularly, if related to modern technical interfaces.

This effect does not necessarily result in degraded cognition, and therefore with appropriate understanding and training, the effects can be to some extent be mitigated.

10.18 SOME SUGGESTED MITIGATING TECHNIQUES

- Avoid taking any action (urge to act) unless obvious.
- Take slow deep breaths and regain control of your time management principles.
- Only perform simple actions.
- If you get stuck trying to understand any factors resulting from the shock, acknowledge it and focus upon something else.
- Consider handing over or swapping tasks/controls to a colleague (breaking a vicious circle).
- If someone else appears stuck, assess their task and consider confidently offering to temporarily take over the task.
- If the feeling arises (and is recognisable), vocalising the fact can help alert your colleague so that they can help.

We observed in the section **Situational Awareness** what are the main components of good

SA. In a startle event, these may well be the main casualties.

11 The Assessment Process

11.1 ASSESSMENT AND BEHAVIOURAL MARKERS

11.1.1 Overview

In any teaching and learning/training environment, one of the fundamental requirements is to have an objective and fair structured method of assessment. This can highlight areas of strength and weakness in the structure and content of the training syllabus and any issues with the instructors or students' performance. By employing appropriate standards or yardsticks to assess, the process will become more successful. The same argument equally applies to understanding human factors. It is essential to have some form of Assessment and Behavioural Marker system. One of the best available is the NOTECHS system which we will introduce next.

11.2 NON-TECHNICAL BEHAVIOURAL MARKERS (NOTECHS)

Non-Technical Behavioural Marker system (**NOTECHS**) is the aviation industries' standard method of evaluating human performance. Full understanding of this system is outside of the scope of this course and falls under the category of CRM training.

It is obvious just how influential are the attitude and behaviour of both instructors and students, which impact upon the success of any training programme. Studying and understanding these attitudes and behaviours use a complex science known as human factors/CRM, as a part of the wider subject, Applied Psychology. In this publication, we can only provide a brief introduction of this subject.

NOTECHS can be broken down into four main categories.

- **Co-Operation**
- **Leadership and Management Skills**
- **Situational Awareness**
- **Decision-Making**

This system considers and categorises behaviours into:

- **Social Skills**—Co-operation and Leadership and Managerial Skills
- **Cognitive Skills**—Situational Awareness and Decision-Making

It further classifies these skills into elements with their associated example behaviours.

DOI: 10.1201/9781032620220-13

Depending on what behaviours can be observed and how they may impact upon flight safety, they can be further graded into five categories:

- **Very poor**
- **Poor**
- **Acceptable**
- **Good**
- **Very good**

In practice, a Remote Pilot cannot fail an assessment by simply displaying poor non-technical skills alone, unless that specific behaviour was the cause of a subsequent technical failure.

Recently, there have been some new developments with an adaptation of the NOTECHS system evolving into HELINOTS (for offshore helicopter operations and another version specifically for the search and rescue services).

Another proposed version of research known as DRONOTS (specifically for the RPAS industry) is under investigation. A table breaking down these areas can be seen next.

CATEGORIES	ELEMENTS	EXAMPLE BEHAVIOURS
CO-OPERATION (Social Skill)	Team building and maintaining. Considering others Supporting others Conflict solving	Establishes atmosphere for open communication and participation. Takes condition of other crew members into account. Helps other crew members in demanding situations. Concentrates on what is right rather than who is right.
LEADERSHIP & MANAGEMENT (Social Skill)	Use of authority and assertiveness Maintaining standards Planning and coordinating Workload management	Takes initiative to ensure involvement and task completion. Intervenes if task completion deviates from standards. Clearly states intentions and goals. Allocates enough time to complete tasks.
SITUATION AWARENESS (Cognitive Skill)	System Awareness Environment Awareness Anticipation	Monitors and reports changes in system states. Collects information about the environment. Identifies possible future problems.

FIGURE 11.1 NOTECHS or Non-Technical Skills or Behavioural Marker checklist.

CATEGORIES	ELEMENTS	EXAMPLE BEHAVIOURS
DECISION-MAKING (Cognitive Skill)	Problem definition/diagnosis Option generation Risk Assessment/option choice. Outcome review	Reviews causal factors with other crew members. • States alternative courses of action. • Asks other crew members for alternatives. Considers and shares risks of alternative courses of action. Checks outcome against a plan.

Very Poor	**Poor**	**Acceptable**	**Good**	**Very Good**
Observed behaviour directly affects Flight Safety.	Observed behaviour in other conditions could affect Flight Safety.	Observed behaviour does not endanger Flight Safety but needs improvement.	Observed behaviour enhances Flight Safety.	Observed behaviour optimally improves Flight Safety and could serve as an example to other Pilots.

FIGURE 11.1 (Continued)

This second section of the table shows the different levels of assessed behavioural competency separated into five categories ranging between Very Poor, Poor, Acceptable, Good, and Very Good.

12 Job-Specific Human Factors

12.1 SITUATIONAL AWARENESS

12.1.1 Overview

Knowing what is going on around us or Situational Awareness (SA) is an essential skill which can be viewed in proportion to our knowledge and experience. Often, SA just being considered as common sense is maybe too simplistic. SA can be developed and improved. By deconstructing some of the factors, it is hoped that a better understanding can be demonstrated. In the context of Human Factors where teams are working closely together, the concept of shared Situational Awareness becomes even more necessary.

12.2 WHAT IS SITUATIONAL AWARENESS?

Situational Awareness is the ability to identify, process, and comprehend the critical elements of information about what is happening to the team with regards to the mission. More simply, it's knowing what is going on around you!

Examples of SA include:

- Knowing what is going on
- A mental model
- A hypothetical state of an individual which continually changes
- SA is very fragile
- There must be an element of attention
- SA is associated with working memory
- For SA to be useful, the mental model must be shared i.e.: Shared Situational Awareness

Dr Mica Endsley, an Engineer and former Chief Scientist of the United States Air Force, proposed the basic theory of Situational Awareness in 1988.

Situational Awareness is generally considered to have three levels which are explained in the next section.

DOI: 10.1201/9781032620220-14

12.3 WHAT ARE THE THREE LEVELS OF SITUATIONAL AWARENESS?

12.3.1 LEVEL 1—Perception of Current Situation

You notice a red light illuminated on the central warning panel.

12.3.2 LEVEL 2—Understanding of Current Situation

12.3.2.1 What Is that Light?

Examples may include:

- Low oil pressure
- Engine of gearbox oil
- High oil temperature

12.3.3 LEVEL 3—Projection of Future Status

- What are the implications of this emergency?
- Is it critical?
- Is it controllable?
- Will I have to land as soon as possible or divert?
- How much time do I have?
- What resources are available?

The three levels of SA are approximately aligned with the central tenet of Threat Error Management (TEM):

- To avoid a threat, one must be able to project ahead.
- To trap an error, one must understand the situation.
- If one does not understand the error, one is more likely to make an error requiring mitigation.

12.4 WHAT IS MY AWARENESS STATE AT THIS MOMENT?

Your ability to cope with any mitigating procedures will be proportional to your current state of awareness.

12.5 SHARED SITUATIONAL AWARENESS

Have I shared my thoughts and ideas with my colleague and obtained and considered his/her opinion? This is one of the most fundamental principles of SA and will be discussed later in this chapter.

Sharing your own perception and discussing appropriate actions with your colleague significantly increases the probability of a better-quality decision.

AWARENESS COLOUR CODE CHART

WHITE
The lowest level: "You are switched off" and unaware of what is going on around you and really not ready for anything. Reasons why one may be in this condition may include sleep, fatigue, stress, or impairment due to drugs or alcohol.
YELLOW
You are alert and aware but calm and relaxed. You are alert to the surroundings (and environment) and to the people who occupy it and their body language. You are alert not paranoid. In this state, it is difficult for someone to surprise you.
ORANGE
A heightened level of awareness. You sense that something is not right. This is the time to evaluate and form a plan. Evasion and diffusion work best here before the next level.
RED
The fight is on! You are taking decisive and immediate action! Recognising attack rituals and set ups helps one avoid this level.

FIGURE 12.1 The Awareness Colour Code indicates four levels of alertness.

12.6 RECOGNISING LOSS OF SITUATIONAL AWARENESS

- Ambiguity: Information from two or more sources do not agree
- Fixation: Focussing on one thing (i.e.: Attention focus/tunnelling)
- Confusion: Uncertainty or bafflement about a situation (often accompanied by anxiety or psychological discomfort)
- Not prioritising the flying task: (everyone is focussed on non-flying activities)
- Everyone heads down: (Look out the window!)
- Not meeting expected fix: (Checkpoint on flight plan or profile, fuel burn, etc.)
- Non-compliance: (Checklists, SOPs—skipping or ignoring)
- Busting limitations: (Minimums, regulatory requirements)
- Cannot resolve discrepancies: (Contradictory data or personal conflicts)
- Not communicating fully and effectively: (Vague or incomplete statements)

How can we mitigate potential errors in Situational Awareness?

- Predetermine crew roles for high-workload phases of flight (Areas of Vulnerability, AOVs)
- Develop a plan and assign responsibilities for handling problems and distractions

- Solicit input from other agencies, including cabin crew, ATC, etc.
- Rotate attention from plane to path to people (aviate, navigate, communicate)
- Monitor and evaluate current status relative to your plan
- Project ahead and consider contingencies
- Focus on the details and scan the big picture
- Create visual and aural reminders of interrupted tasks
- Adhere to SOPs
- Why brief? Preflight, approach brief, post flight de brief, offshore helideck awareness
- Improve monitoring skills.
- Adhere to standardised deviation calls.
- Use of checklists.
- Avoid confirmation bias when using ECLs.
- Discuss practical applications.
- Discuss and raise awareness of issues with new aircraft type.

If you see SA breaking down SPEAK UP!

12.7 THREAT ERROR MANAGEMENT (TEM)

12.7.1 Overview

With many of the basic principles of being Situationally Aware, the concept of Threat Error Management has evolved in proportion to the ability of being able to classify and analyse a critical mass of safety data which has been gathered over many years. Once this data became organised, it showed patterns of cause and effect of events leading to accidents and incidents. In this chapter, we will set out some of the most relevant criteria.

Threats are all around us. When we walk down the street, driving the car, and in almost everything we do. Human beings are fallible and so make many errors. As we described in a previous chapter, these errors can be slips, lapses, or mistakes and sometimes result from deliberate or accidental violations. Many safety critical industries such have over the years amassed vast amounts of data relating to accidents/incidents, their causes, and methods of future mitigation.

The Aviation Industry has evolved with much data obtained from such sources as Flight Data Telemetry, Cockpit Voice Recorders, and more recently Line Orientated Safety Audits (LOSA).

Only relatively recently did we have the facilities with sufficient power and speed to be able to objectively collect, collate, and analyse such stores of data.

As this data became organised, classified, and placed into the correct context, it was possible to make some useful assumptions which ultimately evolved into the discipline of Threat Error Management.

This is now considered an important component tool in recognising and mitigating aviation occurrences.

12.8 THE THREAT ERROR MODEL

There are three basic components of the TEM model from the perspective of flight crews:

- Threats
- Errors
- Undesirable Aircraft State (UAS)

12.9 WHAT ARE THREATS?

Threats are events that occur beyond the influence of the flight crew, increase operational complexity, and which must be managed to maintain the margins of safety.

These can be classified into the following categories with typical examples as explained in the next sections.

12.10 ANTICIPATED

- Weather
- Obstacles
- Congested airport
- Complex standard instrument departures and standard terminal arrival routes

12.11 UNANTICIPATED

- In flight aircraft malfunction
- Unforecast weather
- Automation anomalies

12.12 LATENT

- Organisational changes
- Optical illusions
- Fatigue
- Stress
- Complacency

12.13 WHAT ARE ERRORS?

Errors are defined as actions or inactions by the flight crew that lead to deviations from organisational or flight crew intentions or expectations.

These can be classified into the following categories with typical examples.

12.14 AIRCRAFT HANDLING

- Manual handling
- Flight controls
- Incorrect mode settings with automatics
- Incorrect radio frequency set

12.15 PROCEDURAL

- Incorrect or poor SOPs
- Briefings
- Checklist procedures

12.16 COMMUNICATIONS

- ATC
- Missed calls
- Incorrect phraseology
- Misunderstood communications
- Pilot to Pilot misunderstanding

12.17 WHAT IS AN UNDESIRABLE AIRCRAFT STATE (UAS)?

UASs are flight-crew-induced aircraft position or speed deviations, misapplication of flight controls, or incorrect system configuration, associated with a reduction in margin of safety.

Some typical examples are:

- Unstable approach
- Continued landing after unstable approach
- Unnecessary weather penetration
- Aircraft control (attitude)
- Vertical/lateral speed deviations
- Operation outside aircraft limitations
- Unauthorised airspace penetration
- Continued landing after unstable approach
- Systems
- Engine
- Automation
- Mass and balance

12.18 POTENTIAL OUTCOMES

A UAS is a transitional state whereas an outcome is an end state.

The potential outcome can be categorised as:

- Return to safe operations
- An additional error
- Occurrence (incident/accident)

12.19 MITIGATING COUNTERMEASURES

12.19.1 Planning

Planning is essential for managing anticipated and unanticipated threats and include the following steps:

- Thorough planning/briefings, concise, not rushed and meeting requirements
- Plans/aims and decisions communicated and acknowledged
- Workload assignment roles and responsibilities defined and communicated for normal and non-normal situations
- Contingency management with effective strategies to manage threats to safety
- Threats and their consequences anticipated, and all available resources used to manage threats

12.20 EXECUTION

Execution is essential for error detection and error response and include the following steps:

- Crew members actively monitored and cross check systems and other crew members
- Aircraft position, settings, and crew actions verified
- Operational tasks prioritised and properly managed to handle primary flight duties
- Avoidance of task fixation
- Avoidance of work overload
- Automation properly managed to balance situational and workload requirements
- Automation set-up briefed to other crew members
- Effective recovery techniques from automation anomalies

12.21 REVIEW

Review is essential for managing the changing conditions of a flight and include the following steps:

- Evaluation and modification of plans
- Crew decisions and actions openly analysed to make sure the existing plan was the best

- Crew members ask questions to investigate and/or clarify current plans of action
- Crew members not afraid to express lack of knowledge and "nothing taken for granted"
- Crew members state critical information or solutions with appropriate persistence
- Crew members speak up without hesitation

12.22 FURTHER GUIDANCE: ICAO LOSA DOC 9803

TEM data often suggests that the experience, routine, and a slight degree of complacency can often promote the feeling:

> "That would never happen to me".

Research and data analysis by CASA suggest the following statistics.

12.23 OPERATIONAL THREATS

- Average of 4.5 per flight
- Ten per cent not correctly managed

12.24 OPERATIONAL ERRORS

- Average of three per flight on 80% of flights
- Twenty-five per cent not correctly managed
- Nineteen per cent lead to additional errors which progress to UASs

12.25 UNDESIRABLE AIRCRAFT STATES (UASs)

- Thirty-three per cent of all flights

12.26 MONITORING AND INTERVENTION

12.26.1 Overview

As many operations become more autonomous than manually handled, the operator is relegated to a role which involves more and more solely monitoring duties. This lack of direct hands-on involvement adds a different dimension to the safety of the task. Human beings like to do rather than watch and so are not perfectly matched to this requirement. In this section, we will investigate some of the factors which impact upon our ability to perform a monitoring role and how we can mitigate some of the issues.

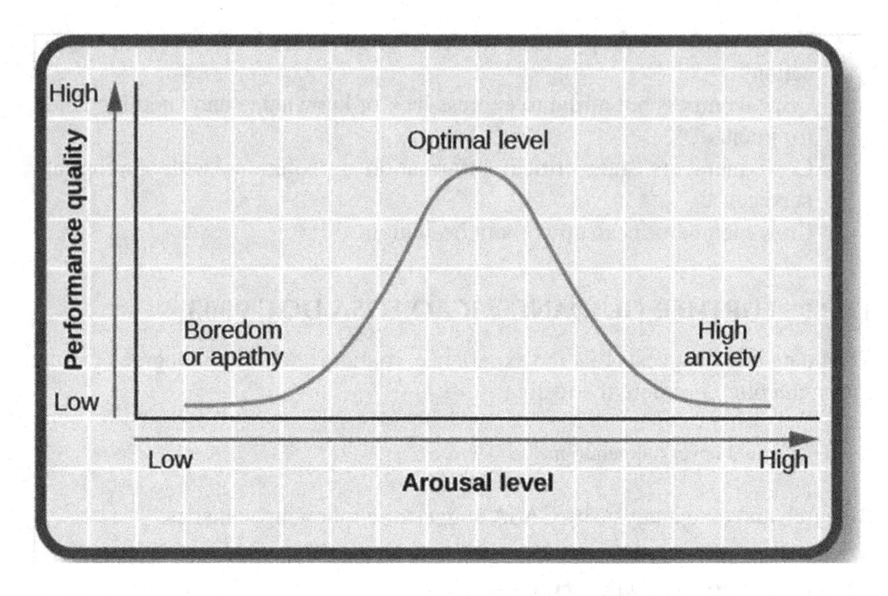

FIGURE 12.2 The Yerkes–Dodson Law of Arousal and Performance.

In many safety critical operations, there will be teams working together where someone will be physically operating equipment while their colleague will be required to monitor their performance, offer appropriate advice, and sometimes take over control in an emergency. This can be a very difficult process for humans since we prefer to be doing rather than monitoring.

In aviation, this task has come under specific research in an attempt to mitigate many of the issues arising from poor monitoring performance in a multi-crew scenario. If we consider the chart below, it suggests that performance quality and arousal (attention) level have a specific relationship according to the subjects' respective levels of boredom or apathy developing into high anxiety at the other end of the scale.

12.27 AREAS OF VULNERABILITY (AOV'S)

One strategy suggested by the **Flight Safety Foundation** is the concept of **Areas of Vulnerability (AOVs).** The basic concept of this guide is to classify a flight into appropriate AREAS OF VULNERABILITY (AOVs) where the risk factor relates to workload and available Pilot capacity to perform tasks dependent upon what the current area of operation demands.

There are three classifications:

- High (RED)
- Medium (YELLOW)
- Low (GREEN)

The following table indicates respective AOVs with corresponding recommendations for maximising the safe conduct of flight.

Level of Vulnerability	Definition		Desired FPM Behaviour		
	In Flight	On Ground	PF/PM	FPM Attention and Sampling Rate	Workload Management Strategy
High (Red Areas)	All changes of: ➢ Lateral trajectory ➢ Vertical trajectory ➢ Speed ➢ Last 1000' of climb or descent ➢ All flight close to ground	Approaching crossing or entering a runway or tight space.	Crew (general) PF PM	Both Pilots maintain total focus on flightpath scan, at a high sampling rate. Undivided attention to flight path Undivided attention to flight path if possible	➢ Avoid any task not related to flight path. ➢ Unavoidable (especially pop up) tasks must be delayed until exiting High AOV or accomplished by PM. Avoid all tasks not related to flight path. ➢ Avoid all non-essential tasks. ➢ Avoid all tasks not related to flight path if possible. ➢ Essential and time-critical tasks (not related to flight path) completed if both brief and unavoidable, but focus must be returned to flight path as soon as possible.
Medium (Yellow Areas)	Climbs and descents. Flights below 10,000'	All other ground movement.	Crew (general) PF PM	At least one Pilot maintains focus on flight path scan, at an elevated sampling rate. Undivided attention to flight path if possible. Flight path is primary, but attention may be divided between flight path and essential tasks.	➢ Avoid any task that is not essential. ➢ Essential tasks may be performed by PM. Keep PF focussed on flight path. ➢ Avoid all non-essential tasks. ➢ Avoid all tasks not related to flight path if possible. ➢ Essential unavoidable tasks requiring PF may only consume very brief moments of attention—return focus to flight path immediately. ➢ Avoid non-essential tasks. ➢ Essential non-time critical tasks (not related to flight path) may be performed but return focus to flight path at frequent intervals.

FIGURE 12.3 The Flight Safety Foundation concept of Areas of Vulnerability (AOVs).

Level of Vulnerability	Definition		Desired FPM Behaviour		
Low (Green Areas)	Straight and level cruise flight above 10,000’	Stopped with parking brake set.	Crew (general) PF PM	At least one Pilot keeps flight path as a top priority but at a normal sampling rate. Flight path is primary, but some division of attention to complete other tasks is permitted. Flight Path is primary, but some division of attention to complete other tasks is permitted.	➢ Manage tasks normally. Tasks not related to flight path preferentially done by PM keeping PF focussed upon flight path. ➢ To the extent practical, use this time to accomplish foreseeable tasks. ➢ Ensure frequent return to flight path. ➢ To the extent practical, use this time to accomplish foreseeable tasks. ➢ Ensure frequent return to Flight path.

FIGURE 12.3 (Continued)

For a full treatise on the concept of AOVs, please follow the link to the publication which can be found in Part Eight—the Resources, Links, and Recommendations.

12.27.1 Final Report of the Active Pilot Monitoring Working Group

A summary of good monitoring practices abstracted from the UK CAA "Monitoring Matters" document:

- Stay in the loop by mentally flying the aircraft even when the autopilot or other Pilot is flying the aircraft.
- Monitor the flight instruments just as you would when you are manually flying the aircraft.
- During briefings, include "monitor me" type comments to encourage intervention—"remind me if I haven't asked for the after-take-off checks".
- Provide the occasional monitoring reminders, e.g.—"make sure that the tail wind doesn't exceed 10 kt".

During the flight, the captain should ensure that the shared mental model remains intact. This can be achieved through:

- Application of TDODAR (Time, Diagnose, Options, Decide, Act/Assign, Review) (agree the plan).
- Expression of intent (I will be flying the descent at 200 kt).
- Providing a situation update to the PM when he/she has been carrying out a non-monitoring task.
- Manage the workload.
- When the workload gets too high, prioritise which parameters to monitor—don't multi-task for too long.
- When dealing with emergency situations, ensure adequate time and space to enable the continuation of the monitoring tasks.
- Avoid programming the FMS at critical phases of flight.
- Mentally rehearse during low periods of workload, monitoring tasks that will occur in the next phase of flight.
- Make cross checking achievement of the autopilot targets a force of habit.
- Verbalise your observations or checklists (especially if a single Pilot).
- At the end of the flight, discuss how well the monitoring was carried out—did you both share the same plan?
- When the aircraft is carrying defects that are acceptable in the MELs, consider the impact on the monitoring task—make a note (mental or otherwise) of the affected flight parameters, modes, or systems that will require more attentive monitoring (discuss this during briefing).
- When referring to charts/checklists/QRH, hold them in a position that facilitates the scanning of flight parameters.
- The PF can put the A/C into a situation where it is unsafe, but PM can stop it "Never whisper when you know it's time to shout".

12.28 AUTOMATION (PHILOSOPHY AND USE)

12.28.1 Overview

This section is written to compliment the previous chapter by suggesting some more practical operational methods of application of the theory which can be highlighted by reflection and self-assessment of your everyday tasks.

Modern technology relies upon automated processes for safe and efficient operation. It can also cause significant incidents when misunderstood or mishandled. In aircraft, misuse may develop into an undesirable state difficult or impossible to recover from using traditional hand-flying techniques. These processes may lead to many instances where the interface and interaction between the human and the machine result from non-technical issues which, if not mitigated by high-quality training, can cause fatal consequences.

One of the biggest problems concurrent with improved automation is the fact that some operators become too reliant on the auto systems, and system knowledge and manual handling skills then develop in inverse proportion to the equipment they are using. Even though exceptionally efficient and reliable in most aviation cases, any system designed and built by a human will at some time probably go wrong! When this happens, the operator often becomes swamped and overloaded with events outside his knowledge, experience, and most importantly his control! An excellent example of this event can be investigated by analysing the 2010 incident when a Rolls-Royce Trent 900 powered Airbus A380–800 being operated by Qantas Airways on a scheduled passenger flight from Singapore to Sydney, Australia, was passing at an altitude of 7,000 feet in the climb in day VMC when an uncontained failure of the no. 2 engine occurred.

This complex emergency is one of the best examples of aviation professionalism where the crew behaviour, time management, and CRM communication exemplified why the highest standards of manual reversion ability and the power of the human mind to synthesise and extrapolate novel solutions to problems need to be current even when operating highly automated equipment (see https://skybrary.aero/accidents-and-incidents/a388-en-route-batam-island-indonesia-2010).

An excellent resource for suggestions as to how best to mitigate automation problems is contained in the **UK CAA CAP 1607 Practical Crew Resource Management (CRM) Standards:**

Some mitigating suggestions include:

- Pilots have in-depth practical knowledge of all auto-flight systems and modes.
- Pilots' use of automation levels is optimal.
- Pilots can move confidently and effectively from one level of automation to another.
- Pilots maintain confidence in handling the aircraft manually.
- Briefings include plans for how automation will be used, and any changes are shared.

- There is active monitoring of auto-flight modes and FMA/FMAs.
- Pilots are confident using the auto-flight system in degraded modes.
- FMS/FMGS programming is included in the briefing and cross-checked.
- Automation errors are spotted and managed in a timely way.
- Autopilot engagement and disengagement are clear to both Pilots.

12.29 GENERAL SELF-ASSESSMENT AUTOMATION QUESTIONS

12.29.1 Use of Automation

- What use is it?
- How do you feel about integrating with automation?
- Does it make your life easier?
- Does it make your life safer?
- How?

12.29.2 Evolution of Automation

- How far can it evolve technically?
- How far do you want it to evolve?
- How has it changed your life as a Pilot?
- How much do you trust automatic systems?

12.29.3 Performance

- How reliable do you find current automated flight systems?
- How easy is it to detect auto system malfunctions?
- Does it reduce your handling skills?
- Does it reduce your monitoring skills?

12.29.4 Complacency

- Are you more complacent with automatics?
- Are you less complacent with automatics?

12.29.5 Features Which Inhibit Understanding

- Describe any features of current automation hardware that inhibit your understanding.
- Describe any features of current automation procedures that inhibit your understanding.

12.29.6 Training

- Do you receive adequate specific automation systems training?
    - What areas of automation training need to be improved?

12.29.7 Knowledge Check

The following questions will give the reader the opportunity to check their progress and understanding of the previous chapter content prior to a review of Learning Outcomes.

1. What percentage of accidents/incidents result from human behaviours?
2. What is the basic principle of CRM?
3. What is the difference between perception and reality?
4. What do you understand by the communications loop?
5. What part of the loop is most often omitted?
6. Explain the difference between a leader and follower.
7. What does the mnemonic PEAR stand for?
8. Using the THERP Workload Management table, which factor most increases the commission of error?
9. Explain confirmation bias.
10. List five environmental (physical) factors impacting stress?
11. List any five common symptoms of stress.
12. Explain a method of mitigating stress.
13. What does the mnemonic OCEAN stand for in terms of personality traits?
14. List the three levels of Situational Awareness.
15. What should you do immediately if your feel you are losing your Situational Awareness?
16. What are the components of Threat Error Management?
17. What do you understand by the term AOV as used in the context of monitoring?
18. Explain the fight or flight reflex.
19. Can resilience be learnt?
20. Explain four factors relating to Performance Adaption.
21. What does the acronym NOTECHS stand for?
22. Explain which of the four categories of NOTECHS' behaviours are described respectively as the social and the cognitive skills?

12.30 LEARNING OUTCOME REVIEW

Before continuing to the next chapter, it is worth reviewing whether or not you believe you have achieved and fully understood the Learning Outcomes from Human Factors chapter.

LO 1: Analyse what factors influence various human behaviours.

LO 2: Investigate appropriate processes to understand, identify, and mitigate those behaviours which affect operational safety.

LO 3: Describe the importance of having a structure allowing the classification and grading of non-technical behaviours.

LO 4: Identify and explain why subjects are described as generic behaviours and job specific.

Part Three

Accident/Incident Investigation/Case Studies

FIGURE PIII.1 Drone accident.

DOI: 10.1201/9781032620220-15

13 Accident/Incident Investigation/ Case Studies

13.1 LEARNING OUTCOMES

By the end of the lesson, students will be able to:

LO 1: Analyse each case study to investigate what factors you think have the biggest influence in causing the occurrence.
LO 2: Identify any common structural or behavioural factors influencing the occurrences.
LO 3: Describe the importance of having appropriate structures to facilitate the identification and classification of outcomes.
LO 4: Identify and explain which factors you feel are the most challenging to isolate.

13.2 OVERVIEW

Case studies are accident or incident reports usually conducted by your own organisation, but when it involves serious injury or loss of life, they are conducted by government agencies. In the case of UK Aviation, they are conducted by the Air Accident Investigation Branch (AAIB).

The process is very comprehensive and thorough and usually comprises reports several hundreds of pages long published in a standardised format.

Due to the size of these reports, it is outside the scope of this publication to reproduce such documents.

Therefore, I have edited several different sample case studies down to a few pages containing the main points to give a flavour of how the larger report would be structured. We will then ask some suitable questions referring to the SMS and CRM principles contained in the previous chapters.

The process of analysis is identical in each case.

13.3 THE STRUCTURE OF ACCIDENT ANALYSIS

A definitive approach to the science of investigation is outside the scope of this publication, and so we are only making assessments in the context of SMS structures and Human Factors influences. For those readers who wish to analyse more case studies,

DOI: 10.1201/9781032620220-16

I have provided links to some of the more frequently used examples I investigate during my classroom training sessions.

I have also included links to some of the more frequently used processes in accident investigations which can be found in **Part Eight—Resources, Links, and Recommendations** including:

Human Factor Analysis and Classification System (HFACS)—Drs Wiegmann and Shappell
Human Factors Intervention Matrix (HFIX) (Shappell & Wiegmann, 2006) analysing:
Bowtie analysis (UK CAA)
Fishbone root cause analysis technique (HSE)

13.4 SOME BASIC PRINCIPLES

When performing your initial analysis, you can only evaluate what you know based upon the evidence of the accident report. There may be other contributory factors, but you must not imagine or assume anything which has not been reported. Personal opinions are not part of a case study. Upon completion, you may allow yourself to form a personal opinion based upon the events you have analysed, but that is all it is—just an opinion.

Being clinical and logical should help to avoid any subjectivity, but even then, different people will come to different conclusions based upon their assessment of event. This actually may not be a bad thing as long as it leads to discussion assisting the formation of improved critical thinking and review.

It is often helpful to start your analysis from the point at which the accident occurred then work backwards listing each critical event as it influenced the final outcome. From then, you can individually analyse each part asking why and listing all your conclusions. By the end of the exercise, you should have a list of influencing factors which can now be revisited and risk assessed retrospectively.

At this point, using the principles you have learnt in the Part One, you should be able to determine many of the structural factors and identify what caused the accident. Once completed, this evidence should enable you to further analyse what behavioural factors contributed to the accident and classify them into a list of main factors. With this information, it should be possible to come to objective conclusions and make recommendations for processes to mitigate a repeat of the events.

Using the guidelines as proposed below:

13.4.1 HFACs

- Unsafe acts of operators (e.g. aircrew)
- Preconditions for unsafe acts
- Unsafe supervision
- Organisational influences

HFIX methodology is designed to develop intervention strategies targeted at preventing or mitigate errors within the four-act-level categories of HFACS:

- Decision errors
- Skill-based errors
- Perceptual errors
- Violations

Five broad areas of intervention, or domains, are proposed:

- Organisational/administrative
- Human/crew
- Technology/engineering
- Task/mission
- Operational/physical environment

Using the HFACS and HFIX criteria, one should suggest asking the questions as listed in the next section.:

13.5 SAFETY MANAGEMENT (PART ONE)

- **Legal**—Are the regulations and standards respected and practised?
- **Moral/Ethical**—Does the organisation practise a Just Culture respecting employees?
- **Structural**—Is there an adequate structure to ensure legal criteria are practised?
- **Educational**—Are there adequate resources and importance given to initial, refresher, and continuation training?
- **Organisational**—How is the organisation governed and operated?
- **Financial**—Are there adequate funds to support the operation, and are they being invested in the correct areas?

13.6 HUMAN FACTORS (PART TWO)

- **Behavioural**—Using a marker system such as NOTECHS, classify the components of the accident as cognitive or social. If they are cognitive, they are relatively easy to mitigate with appropriate training. Problems with Social Skills suggest more complex issues.

Analysing the individual the topics contained in Part Two may guide you towards the most influencing Human Factors relevant to the case study.

13.7 GENERAL QUESTIONS

1. What do you feel are the main contributory factors classifying your findings?
2. What do you think are the event implications for the respective organisations?

3. What do you consider to be the severity of the event by referring to the Risk Severity tables described in the Part One?
4. Suggest mitigating corrective actions.
5. What do you believe is an appropriate timeline for investigation and distribution of results?
6. What do you think you should do with your findings and conclusions?

13.8 EXERCISE

So now we should be able to utilise the lessons we have learnt and apply these same basic principles with any case study we would wish to investigate.

The following four case studies (three aviation related and one marine) should allow you to practise your analysis. Listed as Case study 5 are some tables extracted from the 2019 Airprox Board. For interest, look at the types of incursions and area of incident and think about how to mitigate similar issues in the future.

If you would like to try your new skills further, I have included a list of other case studies/videos which I believe give a broad view of different accidents from different countries with different cultures and organisational structures. Hyperlinks to these case studies can be found in Part Eight. Be aware that some of the videos may be slightly sensationalised for dramatic effect, but the underlying information is relevant.

1. KLM Tenerife
2. US Airways 1549 A320 Hudson River
3. Trans Asia 235
4. Avianca 52
5. United Airlines 232
6. Saudia 163
7. Dan Air DA 1008
8. Teterboro Lear Jet
9. German Wings 9525
10. Air France 447
11. John and the duck

Additionally, I have provided links to the UK Government Air Accident Investigation Branch (AAIB) where you can view various drone accidents found in Part Eight.

NB: Can you notice how many similarities there are between all of the accidents, and what does that tell you?

CASE STUDY 1: AIRBUS A300 B4 ACCIDENT IN OSLO

The purpose of working through this Aviation Ground Operations accident case study is to investigate an actual critical incident which unbelievably did not result in a fatality. However, the Captain who was badly injured suffered permanent physical damage which destroyed his career. It is interesting to see the human cost of this accident and how at a time before SMS and CRM were mandatory legislative requirements, the various parties responsible did not assume any responsibility or accountability until challenged in the Courts of Law.

This incident occurred at a time before SMS was a mandatory legal requirement.

13.9 INCIDENT

Following an emergency landing where an Airbus A300 B4 cargo aircraft returned to Oslo Gardermoen, after an indication of a fire in the cargo compartment, the subsequent investigation indicated that the problem was identified as the synthetic oil used in the Auxiliary Power Unit (APU) situated in the tail section of the aircraft, which had been overfilled with the result that excess oil had dripped into the air conditioning system causing a build-up of gasses smelling of burning plastic. These excess gasses had activated the smoke detector system causing the crew to declare a MAYDAY and return to Oslo airport almost immediately after take-off.

Several days later when maintenance had been completed, the aircraft was towed to the freight area to await loading of cargo for a return flight to Brussels where it was based.

13.10 CARGO-LOADING PROCESS

This process involves a scissor lift equipment positioning at the front cargo door of the aircraft. An LD3 size container is placed upon the lift and then hoisted up in line with the aircraft floor, then manhandled on a series of ball bearing rollers and positioned into the correct part of the cargo cabin.

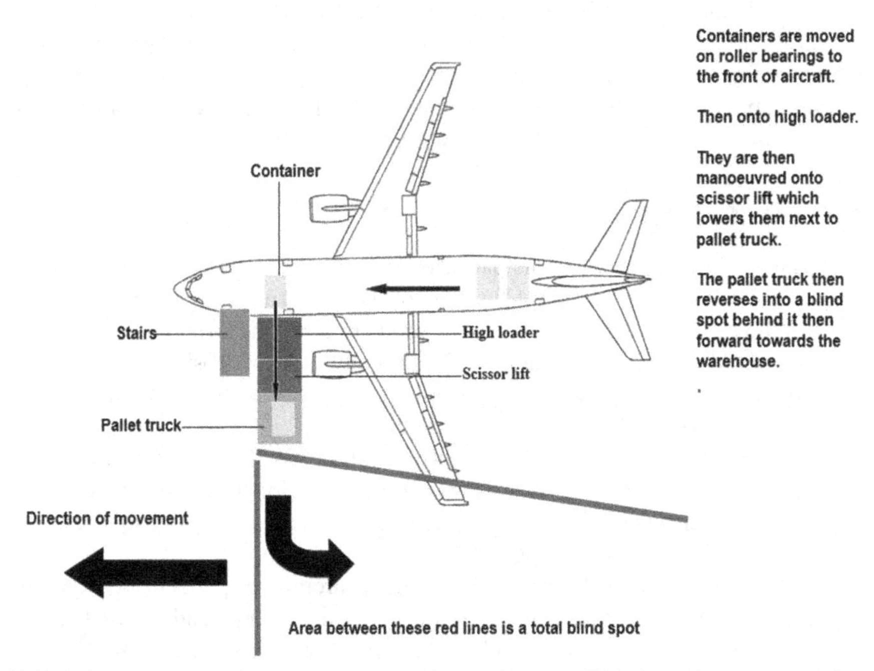

FIGURE 13.1 Loading equipment surrounding an Airbus A300 B4 used in cargo operations.

FIGURE 13.2 LD3 pallet truck.

Normally, the containers are presented to the scissor lift by driving a train of trollies perpendicular to the lift. They are then manually pushed on to it one at a time, then raised up to the aircraft floor. Occasionally, the process is performed by a single pallet/container vehicle replacing the train of trollies and feeding the scissor lift by reversing up to it when the container is manually pushed onto the lift platform. The vehicle then drives away to the cargo depot to collect another container.

13.11 ACCIDENT

1. The 17-year-old driver of the pallet loader had parked his vehicle, so it was completely hidden under the fuselage of the A 300 and was sitting waiting for the arrival of the scissor lift from another part of the airport.
2. The vehicle was tucked under the aircraft belly in such a position that it was not visible to the Captain and the Flight Engineer as they descended the aircraft stairs, until they reached the bottom.
3. The crew intended to carry out an inspection of the tail area of the aircraft containing the APU which had been responsible for their previous emergency landing prior to signing the Technical Log accepting that maintenance had been carried out appropriately.
4. As the Captain walked towards the tail of the aircraft, the pallet loader driver noticed that the scissor lift was approaching the aircraft and realised that he would have to reverse his vehicle and move out of the way to allow it to assume its correct position next to the cargo door.

FIGURE 13.3 Pallet truck from Figure 13.2 but now carrying a standard size LD3 container.

5. Without looking to confirm his path was clear, he reversed his vehicle at high speed to clear the underside of the aircraft and struck the Captain on the left leg with the six-inch diameter steel rollers at the rear of his vehicle.
6. The impact sent the Captain flying into the air and landing on the apron behind the pallet loader. It was only the fact that the Flight Engineer who witnessed the accident and screamed for the driver to stop that the Captain was not run over and crushed by the vehicle.
7. The Captain's next recollection after regaining consciousness was being placed into a vehicle and driven to a hospital nearby.

13.12 POST-ACCIDENT INVESTIGATION

To this day, to the best of his knowledge, there was no official accident investigation which included the Captain's or indeed any other participant's or observer's testimonies.

The Captain received a five-minute phone call from a member of staff who did not identify their position about six weeks later. He was not allowed to see any documentation relating to the accident if it did actually exist.

Ironically, about eight weeks later, the Captain, with slowly deteriorating physical ability, was sent to Oslo to perform the same series of flights. He conducted his own investigation which formed the basis of a later court case in Norway. It took three years for the company with oversight responsibility to accept liability and a further three years before a subsidiary company settled out of court. As part of his following year-long investigation including a DVD of the whole operation he made on his return visit, the captain concluded his own contributary findings which are listed in the next section.

By now, the spinal injuries had deteriorated to the extent that the Captain was no longer able to walk without crutches and had his Medical Certificate suspended. The company refused to pay his salary for nearly 18 months, and they also prevented the payment of his loss of licence insurance.

13.13 POST-ACCIDENT FINDINGS

1. The driver was a student working part time during an Easter Break.
2. He did not receive any formal driver training for his role at the airport.
3. He did not look before reversing his vehicle.
4. The vehicle was not fitted with adequate rear facing or wing mirrors.
5. The reversing warning beepers were unserviceable and had been for many months.
6. All drivers routinely exceeded the apron speed limit of 5 mph. An example of this behaviour was proven by the video footage taken by the captain at a later date. (The apron is constructed using 10-metre square concrete sections surrounded by a black bitumen expansion joint.) By comparing the real-time video footage indicated on the DVD camera screen related to the number of concrete squares covered by the pallet trucks, it was estimated that regular speeds of up to 25 mph were routinely conducted by the various drivers.
7. It was discovered that the drivers were under severe commercial pressure to load as many pallets onto the scissor lift as quickly as they could.

13.14 ADDITIONAL INFORMATION

- There was no clear policy in place clearly defining who had oversight responsibility and accountability in this type of operation.
- The captain was employed by a UK agency based in London but registered in Jersey.
- He was subcontracted to a Belgian freight operator who employed 50% Belgian aircrew and 50% contractors, but no one accepted or took direct overall responsibility for the operations within the network.
- The Belgian freight operator was operating on contract with one of the world's largest US freight forwarders, but it appeared that different legal obligations applied in Europe.
- The company responsible for the freight operation in Norway was subcontracted to the Belgian carrier and had just been purchased by the Norwegian state operator.
- There was no formal standardisation of operations and apparent oversight from any of the respective companies involved in this process, and

each in turn delayed any investigation process by claiming it was not their responsibility.
- There was no provision for dealing with an accident on the apron and no first aiders present.

Finally, no formal investigation was conducted by any of the stakeholders.

13.15 OUTCOMES

The captain suffered progressive reaction to significant nerve damage which over the next few months rendered him to lose all feeling in the left side of his body for over one year. This was only partially restored after over a further year's exercise and physiotherapy treatment.

Interestingly, the company refused to pay any salary during the recovery period. When he was later invited to renew his licences by undertaking one week in the simulator, the following events occurred:

> During the de-brief following the final flight test conducted by the company Chief Pilot, the crew were verbally congratulated on successfully completing the test to the company standard. The Chief Pilot then signed the licences of the first officer and the flight engineer then asked to speak privately to the captain. During this exchange, he informed the captain that he would be unable to sign off his licence unless he signed a nondisclosure agreement relating to his accident in Oslo and a separate document stating that he no longer had any legal claim against his employers! After explaining to the Chief Pilot that he would report him to the Belgian and UK Aviation authorities since he had already verbally confirmed that ALL the crew performed to the test standards. He then signed the licence!

Later that week, the captain was informed that since he had been absent from flying for so long, he would be required to be reassessed as competent by undertaking an indeterminate period of line training up to maybe 100 sectors (about six months). The initial training he had received during type conversion only amounted to a maximum of 40 sectors. He was informed that no salary would be paid until successful completion of this training.

13.16 EXERCISE

1. Follow the sequential process of obtaining and evaluating the initial incident report as should be documented in Q Pulse or similar tools. (In this case, there was no formal company investigation process involving the Captain and the only facts are his own investigation).
2. Assuming the roles of a triage team (had one been available), make your recommendations to the Safety Review Board for further mitigating actions.
3. Assuming the roles of members of the Safety Action Group, make recommendations for corrective actions.
4. Discuss your follow-up actions and what you consider to be an appropriate timeline.

5. Do you think it will be necessary to involve a formal Accident/Incident Investigation Team?
6. Discuss what you consider to be suitable recommendations to ensure appropriate mitigations.
7. Do you think the group of organisations responsible followed what we now know as a Just Culture?
8. List what you consider to be the most relevant factors in:
   a) Post-accident analysis.
   b) Mitigating lessons to be learnt.

13.17 KNOWLEDGE CHECK

The following questions will give the reader the opportunity to check their progress and understanding of the previous chapter content prior to a review of learning outcomes.

The science of analysis of case studies follows quite strict guidelines; however, the solutions and opinions may often be quite subjective, and different assessors may often come to quite different conclusions.

Please use your new knowledge from the previous Parts One and Two to analyse then answer the following questions. I have included my own personal opinion following the questions for comparison. I do not claim this is the definitive correct answer but simply my opinion based upon my own personal career experiences.

Case Study 1 Questions

1. What are the most safety critical issues you have identified in the case study?
2. List the different factors you have identified.
3. Describe which issues you consider as structural.
4. Describe which issues you consider as behavioural.
5. Describe which issues you consider as cultural.
6. Describe which issues you consider as financial.
7. Describe which issues you consider as professional.
8. Describe which issues you consider as legal.
9. Describe which issues you consider as moral.
10. Describe which issues you consider as ethical.
11. Describe which issues you consider regulatory.
12. Analyse what counter measures you think would mitigate the most common issues.
13. Explain your opinion if safety procedures have improved over the past years.
14. If your answer is yes to the above question, describe those improvements.
15. Identify which factors you consider as common denominators in the highlighted case studies.

CASE STUDY 1 AUTHOR OPINION (ANSWERS)

1. The accident occurred at a time (2000) when there were no mandatory Safety Management structures and procedures legally required and enforced. Absence of these processes allowed organisations to follow random investigations often conducted by unqualified persons with the brief being to reduce any financial expose arising from any litigation as the priority. So, the lack of legal structures was probably the main influence not helped by the driver's poor apron discipline resulting from poor training and lack of experience for such a responsible position.
2. Answer largely as above. No Risk Assessment at planning stage. Commercial pressure impacting safety procedures. Poor training.
3. The general unprofessionalism of the operation where outcomes without structure will be statistically more random and potentially more serious.
4. The driver's lack of situational awareness. Compare with information contained in the appropriate Human Factors section.
5. In those days, cultural apathy, laxity, and lack of respect for employee's safety were common. Lack of clear chain of command and operational oversight indicated an uncaring culture. Many corporations cynically base their operations in countries where they may take advantage of that specific country's cultural norms such as respect for the law or flexibility of interpretation of the law. Liberia was often used to register shipping companies, and Belgium was often used in aviation because of the ease of obtaining legal derogations from their Aviation Authority.
6. Obviously, businesses need to make a working profit, but many in aviation did and still do this at the expense of cutting corners. Often with shareholder profit being the only consideration; in particular, when often the business model is to maximise short-term profit and take the risk that nothing will happen allowing the business to be sold on. This is indeed a massive risk. Because if you cut corners with safety training then have a major accident, see how the costs will rapidly escalate wiping out any profit you had earned.
7. The whole operation was unprofessional at just about every level and relied upon good luck and not good management. Analysis of Part One shows how simple but basic safety management principles can drastically reduce any exposure to risk. Lack of any Quality Assurance programme and obviously few objective auditing and review of procedures contributed.
8. The legal process was greatly inhibited because of the lack of transparency of any clear chain of command indicating who was actually accountable. Many multi-international corporations hide behind complex contracts using many subcontractors to enable any problems to be directed away from their own culpability.
9. Sadly, moral and ethical considerations as suggested in the next question are often ignored. A clear Mission Statement or "Bill of Rights" which are now integral components of the Accountable Manager's statement as legal requirements of a Safety Management System are included specifically to protect employee's rights.

10. As above.
11. Analysis of the principles of Threat Error Management indicates that a large proportion of threats may be predicted and mitigated by having clear organisational oversight. Regulatory oversight has reduced in many countries as their organisations are privatised and need to justify a profit. Often, the rules are well written and fit for purpose, but the policing is poor with investigations taking too long and those truly responsible being immune from punishment.
12. The introduction of clear standardised SMS requirements as a pre-condition to authorise organisations to conduct their business can mitigate most issues if understood and implemented. However, before you get the opinion that all fault lies with management, this is not true. The effectivity of an SMS also depends upon everyone in the organisation understanding what it does and how it works. But it is essential that everyone contributes to ensuring the data inputs and running of the system, and this needs input from everyone.
13. Safety procedures have improved vastly in recent years, but often aviation is a classic example of how complacency can ruin a good system. That is precisely why the Safety Cycle is so important—continuous monitoring, continuous review, continuous update, and continuous training. However, that relies upon everyone in an organisation taking an active part in the process.
14. Answered in previous question.
15. After analysis of all the case studies in this publication and years of personal investigations, my opinion is that most serious incidents can be risk assessed and mitigations put in place if we follow the rules of SMS and understand the principles of the generic and job-specific Human Factors as advocated in this book.

13.18 COMMENT

As a postscript to this investigation, the treatment of the captain by the company was diabolical. One aspect which indicates the moral and ethical dimension of the responsible company is what happened to his loss of licence payments.

Once the captain's Pilot licence was suspended due to problems walking, the contract specified that after 90 days, a payment of £100,000 was due assuming he would not be able to return to work for a period of up to one year.

After about six months' wait and still walking with crutches, the company phoned to say that they hoped he would return to work eventually, but meanwhile he would have to travel to Brussels to attend a Dangerous Goods course to keep his currency. It would only take one day, and they would pay for the transport from UK to Brussels.

He attended, passed with 98%, signed the course completion certificate, and was back home in the UK later that day.

When it became obvious that it would take several more months to regain his Medical Certificate and with no salary paid, the captain realised that the 90-day period qualifying for the loss of licence insurance payment had elapsed.

He duly applied and four weeks later received a letter from the insurance company saying his claim was invalid. When he phoned to get an explanation, he was informed that since he had returned to work with the 90-day qualifying period, he was no longer entitled to the £100,000 and would have to make a secondary claim after a further 90-day period. When he contested this stating that he did not have a valid medical certificate and could not fly and by definition was unable to work, the reply came:

> On xx.xx.xxxx you attended a Dangerous Goods Course in Brussels passed with 98% and signed the course attendance sheet so technically you have worked and that invalidates your primary claim, You will need to wait a further 90 days and apply for a secondary claim.

He did this and received a cheque next week for £ 9000 ! When contesting this amount, he was told if you read the small print it states that a secondary claim is £9000 and a third claim is £ 300 !

Surprised at how this information came to the attention of the insurance company, it later transpired that the Airline Accountable manager was the husband of the Chief Executive of the Belgian Insurance Company (which is still trading)!

The Accountable Manager is now serving a long sentence in jail in Brussels for fraud for numerous offences.

At the time of writing this book in the UK, there are two high profile cases, the Post Office and the Contaminated Blood scandal which clearly illustrate how a lack of transparent processes including adequate grievance procedures can be manipulated by those in power who although technically accountable hide behind a wall of bureaucracy trying to avoid their legal responsibilities.

The SMS processes which are a legal requirement in aviation now greatly reduce the chance of avoiding this type of outcome.

CASE STUDY 2—S 92 HELICOPTER TAIL ROTOR CONNECTING CABLE INCORRECTLY FITTED

Overview

The purpose of working through this Aviation Maintenance case study is to investigate an actual critical incident which unbelievably did not result in a fatal accident.

NB:

This incident occurred at a time when it WAS a mandatory legal requirement for the company to operate in accordance with their SMS principles in order to be allowed an Air Operators Certificate (AOC) in compliance with EASA ORO. GEN.200.

This process, by applying best practice SMS principles, should allow us to identify many of the respective weaknesses and failures of this organisation's SMS and inherent toxic safety culture.

It should also highlight that functional, robust SMS practices can *never* be allowed to become just a box-ticking exercise. They depend upon all organisation members understanding, reporting, and contributing to the process in an honest way in the context of a *Just Culture*.

13.19 INCIDENT

During a routine 500-hour maintenance check, it was discovered that the cable connecting the foot pedals of an S 92 to the mechanism controlling the pitch of the tail rotor blades had been incorrectly fitted.

Upon further investigation of maintenance records, it was established that this had in fact been the situation for the previous 1,500 hours of operation!

During the 500-hour maintenance check, it is required to check the condition of the cable connecting the control pedals and the tail rotor pitch control mechanism.

This requires the cable to be de-tensioned and the connector nipple to be removed from the slot housing it. Once this condition is approved, the reverse operation is performed and the cable re-tensioned using the approved torquemeter pre-tensioning tool (turnbuckle).

The maintenance manual clearly states that this is a *Critical Maintenance Operation* requiring to be cross-checked by another qualified engineer. The process is known as **a** *Duplicate Inspection*.

13.19.1 Inspection

13.9.1.1 Incident Report

After the incident had been triaged by the Safety Committee and assigned the highest classification of risk, it was decided to order a full accident investigation. The lead accident/incident investigator was tasked with a colleague to perform an appropriate research. This process, which took six weeks to complete, required:

1. Interview of all contributors.
2. Study of all the appropriate maintenance manuals.
3. Analysis of any technical and non-technical contributory factors.
4. Writing a report of findings, corrective actions, and any recommendations.
5. Presenting this report to the Head of Department who would decide upon what follow-up actions would be implemented.

13.20 FINDINGS

The following indicate an abridged list of the more significant findings:

1. The schedule of the maintenance was performed by two different teams of engineers due to time pressures.
2. There was no time allocated for a comprehensive handover between the two teams.
3. Appropriate updated and revised maintenance manuals were not always available.
4. The fact that the process was classed as a CRITICAL MAINTENANCE TASK was not appreciated and understood by some of the engineers.
5. Duplicate inspection procedures were not respected.

6. An appropriate tensioning tool was not available to the first team as it was missing from the tool store even though records show that the tool was stored there.
7. The tensioning process was delayed by over 90 minutes while an engineer from Team A had to go and borrow a tool from a competitor based on the other side of the airport.

This coincided with the end of his shift and transfer of the task to Team B.

8. When replacing the cable, the connector was incorrectly fitted, forcing it into the housing instead of ensuring that it was correctly fitted into the slot and then tensioned. Friction was the only reason the connector did not come away from its housing.
9. Flying hours numbering 1,500 had elapsed since records confirmed later that during the next inspection, the process was obviously not even performed (but signed off) since the incorrect fitting of the connector had not been corrected. **(This factor was one of the most incredible aspects of the investigation and hard to believe that the helicopter had performed 1,500 hours of flight in this state).**
10. The mechanic responsible later admitted that he overlooked the process because of commercial pressure since the company was operating with between two and three engineers short of the required complement for each shift. This was a common practice. The Chief Engineer had documentary evidence showing a routine shortfall of suitably experienced and qualified mechanics on most of the shift plans.
11. Due to time pressure, the company routinely did not allocate sufficient time at the end of shift to clean, replace, and record the return of tools and equipment in appropriate storage facilities.
12. There was a toxic culture prevailing in this department with two engineers refusing to work together due to perceived lack of respect.
13. The basic problem involved cultural differences about disagreement over professional standards due to over 40 years' age difference between shift supervisors.
14. The whole department was operating in a climate of fear and bullying from senior management.

13.21 OUTCOMES

The Director of Engineering made a formal complaint to the Quality and Safety Department about the investigation team, principally the lead investigator, because the practices discovered during the investigation reflected so badly upon his leadership, and they did not single out any engineering staff member to fire! He cited that the Lead Investigator had obviously not been trained correctly and had taken too long to publish his findings (six weeks).

The Lead Investigator was not asked to perform any more Accident/Incident investigations and never found out what if any mitigating decisions were made to future operations!

13.22 EXERCISE

Follow the sequential process of obtaining and evaluating the initial incident report as documented in Q Pulse.

1. Assuming the roles of the triage team, make your recommendations to the Safety Review Board for further mitigating actions.
2. Assuming the roles of members of the Safety Action Group, make recommendations for corrective actions.
3. Discuss your follow-up actions and what you consider to be an appropriate timeline.
4. Do you think it will be necessary to involve a formal Accident/Incident Investigation Team?
5. Discuss what you consider to be suitable recommendations to ensure appropriate mitigations.

13.23 KNOWLEDGE CHECK

The following questions will give the reader the opportunity to check their progress and understanding of the previous chapter content prior to a review of learning outcomes. The science of analysis of case studies follows quite strict guidelines; however, the solutions and opinions may often be quite subjective, and different assessors may often come to quite different conclusions.

Please use your new knowledge from the previous Parts One and Two to analyse and then answer the following questions. I have included my own personal opinion following the questions for comparison. I do not claim this is the definitive correct answer but simply my opinion based upon my own personal career experiences.

Case Study 2 Questions

1. What are the most safety critical issues you have identified in the case study?
2. List the different human factors you have identified.
3. Describe which issues you consider as structural.
4. Describe which issues you consider as behavioural.
5. Describe which issues you consider as cultural.
6. Describe which issues you consider as financial.
7. Describe which issues you consider as professional.
8. Describe which issues you consider as legal.
9. Describe which issues you consider as moral.
10. Describe which issues you consider as ethical.
11. Describe which issues you consider regulatory.
12. Analyse what counter measures you think would mitigate the most common issues.

CASE STUDY 2 AUTHOR OPINION (ANSWERS)

1. The fact that even though there are legally required processes to be obeyed, they were obviously not. Why?
    - Failure to perform duties professionally
    - Failure to read the Maintenance Manual
    - Failure to follow Duplicate Inspection Procedures
    - Not organising the timeline to perform critical tasks with the same crew when possible.
    - Tools not being where they should be and incorrectly recorded in the tool control log.
2. Poor Just Culture starting with senior management to shop floor supervisors.
    - Engineers reluctant to resolve disputes
    - Generational differences over work practices
    - Workload issues
    - Insufficient trained staff
    - Poor oversight
3. The structural regulations were in place but not being observed. How is that possible?
4. There are many behavioural issues:
    - Poor communication
    - Poor leadership and management
    - Ineffective conflict management
    - Poor workload management
    - Personality, cultural and generational differences.
    - Poor stress management.
5. The main cultural issue is the fear of speaking up and letting problems escalate and fester.
6. There may be financial considerations. Why was the team understaffed?
7. The factors highlighted in Question no. 1 indicate a culture where a lack of professional behaviour is either not recognised or recognised but not mitigated, and poor discipline is the outcome.
8. As discussed, this incident is not in legal compliance with the regulations relating to common aviation engineering practice.
9. It is not morally right to expect staff to perform contrary to regulations or be aware of their non-conformance.
10. The company mission statement signed by the Accountable Manager as a pre-condition to being allowed a permit from the CAA to operate as a commercial entity with fare-paying passengers is not being respected.
11. As above, many of the practices such as failing to perform a duplicate inspection which should be understood by all are not respected.
12. Much more frequent and rigorous auditing of processes and practices conducted by organisation's Quality Assurance department and regulatory bodies. Much more stringent sanctions placed upon nonconformities not being mitigated.

13.24 COMMENT

This incident occurred at a time when SMS and Human Factors training were mandatory requirements for commercial operators.

Why do you think it was still possible for so many errors to go unnoticed, and so many processes and individual's performances to circumvent the legal requirements?

It is interesting and maybe prophetic that the main comment from the Head of Department asked of the Accident Investigator was: "What type of poor investigation is this? You have not told me who to sack? Why not?"

This is directly contradictory to the basic CRM concept of what is really important is:

WHAT IS WRONG NOT WHO IS WRONG.

CASE STUDY 3—INCIDENT REPORTING FLIGHT DECK CRM ISSUES

This main case study has been included since it was used as a discussion document during a CRM Training session in order to assess respective crew attitudes to using mobile phones during a flight and gives an interesting perspective on why reports need to be submitted if an occurrence is considered to impact upon operational safety. The report has been deidentified and appears word for word as submitted by a helicopter captain employed by a Northern European Organisation of which he is a national.

A second much shorter case study is included after as it illustrates a very similar attitude to operational practice and poses the question:

> What types of behaviour do employees feel is culturally, socially, and professionally acceptable and compare the result with what is actually practically required to perform tasks safely? The contents are good examples of the type of discussions which reports should highlight and then be discussed by the Flight Safety Committee as later highlighted in **Chapter 18—Organising a Flight Safety Meeting.**

13.24.1 CRM During Two Consecutive Nights Working Together

Due to OMA regulation 11.2.1, all reports have to be written in English. Therefore, I am writing this report in English.

At the beginning of this report, I would like to emphasise that my intention is not to harm or criticise another colleague in public. This has never been my purpose. I prefer to communicate directly within and after my flights with my teammates.

I was very astonished to find out that my Co-Pilot is not sharing the same values. Unfortunately, she was not satisfied with my performance after two-night flights, and she reported it to the base chief a few days later. She never mentioned anything after the first or even second flight directly to me. During the flights, I gave her a debrief on her approaches to land and some suggestions for improvements. A detailed debrief was not possible because she left the base approximately ten minutes after landing.

I was called into report to my base chief one week later on the xx.xx.xxx.

Therefore, I am writing this occurrence as it happened.

I believe that we do have a CRM culture in this company (in most cases) and are using a briefing and, even more important, a thorough debriefing.

If something goes wrong during a flight or a crew member is not happy about a certain procedure or behaviour, it has to be discussed within the crew immediately after a flight in order to find a solution that everyone agrees on (OMA 8.3.17.4).

If this is not successful, you can go further.

This was not done on flights XXXxxx and YYYyyy on the xx.xx.xxxx.

I have been reported among other things for approaching the rigs too fast. These aspects have never been pointed out to me during, on, or after the flights.

My impression of these two flights are as explained in the next sections:

13.24.1.1 Flight XXXxxx

- The Co-Pilot showed up just 45 minutes prior to the flight, even if it was the first day at the working period and there have been marginal weather conditions at night.
- There seemed to be no real interest in planning the flight together.
- No synchronised iPad from the Co-Pilot during start up in the cockpit.
- Missing steps (floats armed) during final checks.
- Nearly a white out in snow during the landing at XXXX (which had been debriefed).
- There was no comment that she was "scared" during the rig approach. We have been talking that it was a little bit too fast but safe.
- A thorough debrief from the Co-Pilot after the first flight about her feelings and concerns would have been appropriate.

13.24.1.2 Flight YYYyyy

- The Co-Pilot was more concentrated this time.
- But still checking the SMSs on her mobile phone after the level off checks.
- I was using the same safe technique for the two rig landings. It was night but clear sky conditions offshore.
- Again, I was not aware of any concerns of my Co-Pilot.
- Approaching XXXX via ILS 07 and Co-Pilot landing in a snow shower.
- The Co-Pilot was very quiet during the approach and came nearly to a hover (20–30 knots ground speed at the threshold).
- Hovering taxi along the runway until taxiway X and again nearly coming to a white out, I was guiding her through the landing and pointing out the altitude in order to avoid the white-out conditions.
- Since she was still very hooked and quiet during taxi, I asked if I should take controls, which was denied.
- The Co-Pilot left the base approximately ten minutes after the landing without a debriefing.

This leads to the following questions:

- Why is a report written without members talking to each other?
- Why was the Co-Pilot not pointing out her concerns and using normal debriefing procedure using CRM techniques?
- Or is there another motivation goal behind it?

My intention of writing this Occurrence Report is to encourage every colleague, technician, captain, and Co-Pilot to practise a reasonable CRM culture and talk to each other in order to avoid incidents like this. Finally, I am deeply disappointed to be forced to use this way of communication.

13.24.2 Senior Flight Instructor CRM While Supervising a Line Training Check

A Senior Line Training Captain was conducting a Line Check on another very experienced Captain in Dammam, Saudi Arabia. While flying in the Terminal Control Area and performing the PM (Pilot monitoring) duties, the Senior Pilot missed several radio calls from ATC. The captain under test acknowledged these missed calls as he continued with his PF (Pilot Flying) duties.

The PF noted that the Senior Training Captain was rhythmically tapping his hands and feet and shaking his head back and forth during these missed calls.

Once they had exited the Dammam TMA, the Captain under test asked his instructor if he felt fine because of his tapping behaviour and the fact that he had missed several radio calls.

The instructor smiled and explained that he had just bought a brand-new headset that allowed him to Bluetooth his music collection from his iPad to his headset while monitoring the frequency. He said it was very clever and cut out the music to prioritise RT calls when they were transmitted while still being able to listen to AC/DC! Obviously, it did not work correctly.

13.25 ASSESSMENT OF BEHAVIOURS AND CULTURE

The analysis of such case studies usually indicates that 70 to 80% of accidents/incidents are caused by human non-technical attitudes and behaviours.

13.26 EXERCISE

Try to identify what hazardous attitudes and behaviours you notice that are common to both studies.

To help your analysis, refer to the following table indicating five common hazardous attitudes.

Anti-Authority	Anti-authority means a resistance to authority or rules. Pilots with this mindset might say, "Don't tell me what to do!" or dismiss regulations as unnecessary. They may also rationalise not following rules due to exceptional circumstances. However, it is crucial to remember that rules exist for safety reasons and bending them can lead to accidents. Always respect authority and adhere to established procedures.
Impulsivity	Impulsive persons tend to rush decisions without considering all available information. When faced with time pressure, they might act hastily. The key is to pause, assess the situation, and avoid making snap judgements. Taking time to think before acting is essential for safety.

Invulnerability	Those with an invulnerability attitude underestimate risks and believe accidents won't happen to them. They might think, "It won't happen to me; I'm invincible". Acknowledging vulnerability and recognising that safety precautions apply to everyone are crucial. Complacency can lead to dangerous situations.
Macho	The macho attitude involves overconfidence and a desire to prove oneself. Those with this mindset might take unnecessary risks to demonstrate their skills. Remember that even experienced people can make mistakes. Humility and adherence to safety protocols are essential.
Resignation	Those experiencing resignation may feel overwhelmed or defeated. They might think, "What's the use? Nothing will change". This defeatist mindset can lead to complacency and a lack of vigilance. It is essential to maintain a continuous focus on safety and seek solutions rather than giving in to resignation. Did you observe any of the above attitudes?

13.27 PERSONALITY, CULTURAL AND GENERATIONAL DIFFERENCES, AND MONITORING AND INTERVENTION

13.27.1 Knowledge Check

The following questions will give the reader the opportunity to check their progress and understanding of the previous chapter content prior to a review of learning outcomes.

The science of analysis of case studies follows quite strict guidelines; however, the solutions and opinions may often be quite subjective, and different assessors may often come to quite different conclusions.

Please use your new knowledge from the previous Parts One and Two to analyse and then answer the following questions. I have included my own personal opinion following the questions for comparison. I do not claim these are the definitive correct answers but simply my opinion based upon my own personal career experiences.

CASE STUDY 3 (2) QUESTIONS

1. How do I view my role when I am operating in a Pilot Monitoring role?
2. Is it acceptable to read a newspaper in the cruise?
3. Do I ever check my phone or send messages when I am operating?
4. Do I own a Bluetooth Headset that allows me to stream music from my phone while monitoring the radio or performing other safety critical duties?
5. If I want to read a document during operations, do I inform my colleague?
6. Should I inform him?
7. Do I engage in a true flight following mind map when assigned Pilot Monitoring duties or do I only engage during what I consider to be the critical phases of the flight?
8. Which phases of flight do I consider to be non-critical?

9. Do you alter your instrument scan rate significantly as appropriate to the phase of flight Area of Vulnerability (AOV)?
10. What are the most safety critical issues you have identified in the case study 3 (02)?
11. Describe the issues you consider as structural.
12. Describe the issues you consider as behavioural.
13. Describe the issues you consider as cultural.
14. Describe the issues you consider as financial.
15. Describe the issues you consider as professional.
16. Describe the issues you consider as legal.

CASE STUDY 3 AUTHOR OPINION (ANSWERS)

1. As an integral part of the crew just as important as if I am the handling Pilot.
2. No.
3. No.
4. No, this is not professional.
5. It is best practice to keep your colleague in the loop at all times.
6. Yes.
7. It is best practice to keep in touch with the big picture all the time and treat all aspects of flight with appropriate concentration.
8. I treat all phases as critical enough to devote appropriate focus.
9. The scan rate should increase in proportion to the complexity of the phase of flight.
10. The Training Captain attitude was not compatible with his role. He should be leading by setting a good example.
11. This appears to be an example of individual anti-authority behaviour displayed by the Training Captain which could also indicate that the Training Department have some issues with selecting their officers.
12. As above but an indication of a macho attitude that "I know best"!
13. May be cultural if others display the same disregard for the regulations.
14. Not applicable.
15. Not acceptable or professional.
16. Illegal since in contravention of normally acceptable Operations Manual regulations.

13.28 COMMENT

With reference to Case Study 3 (2), I once asked a class of about 12 Pilots their opinion about the Co-Pilot using a mobile phone during a critical phase of flight. I was quite surprised at how many including one training captain (about 33%) did not think it to be a problem, whereas most of the class considered it very unprofessional and even a hazardous attitude. Interestingly, those who did not object were the younger members of the group.

This is a clear indication of how cultural and generation differences of culture and opinion may often conflict.

CASE STUDY 4—HERALD OF FREE ENTERPRISE

This extract is from **DEPARTMENT OF TRANSPORT THE MERCHANT SHIPPING ACT 1894 mv HERALD OF FREE ENTERPRISE Report of Court No. 8074. Formal Investigation**

Over 190 people died when the roll-on roll-off ferry capsized off Zeebrugge, Belgium, on 6th March 1987. The bow doors had been left open after departure, and water flooded the car decks.

The tragedy led to new safety regulations in the British ferry industry. With the sinking of the Estonia for similar reasons in 1994 and the deaths of 850 people, new rules in international ferry safety were introduced in 1999.

The Crown Prosecution Service charged P&O European Ferries with corporate manslaughter in 1989 and seven employees with manslaughter.

The case collapsed but it set a precedent for corporate manslaughter being legally admissible in an English court.

The immediate causes of the disaster were as follows.

1.1 The Herald capsized because she went to sea with her inner and outer bow doors open.

From the outset, Mr. Mark Victor Stanley, who was the assistant bosun, has accepted that it was his duty to close the bow doors at the time of departure from Zeebrugge and that he failed to carry out this duty. Mr. Stanley had opened the bow doors on arrival in Zeebrugge.

Thereafter, he was engaged in supervising members of the crew in maintenance and cleaning the ship until he was released from work by the bosun, Mr. Ayling. Mr. Stanley then went to his cabin, where he fell asleep and was not awakened by the call "Harbour Stations", which was given over the Tannoy address system. He remained asleep on his bunk until he was thrown out of it when the Herald began to capsize. Mr. Stanley has frankly recognised his failure to turn up for duty, and he will, no doubt, suffer remorse for a long time to come. If the company regards it as appropriate or necessary to take disciplinary action against Mr. Stanley, it has power to do so under the Code of Conduct for the Merchant Navy. In fairness to Mr. Stanley, it is right to record that after the Herald capsized, he found his way out of the ship on to her hull where he set about rescuing passengers trapped inside. He broke a window for access, and, when he was scooping the glass away, his right forearm was deeply cut. Nevertheless, he re-entered the hull and went into the water to assist passengers. He continued until he was overcome by the cold and bleeding.

1.2 The bosun, Mr. Terence Ayling, told the court that he thought he was the last man to leave G deck, where he had been working in the vicinity of the bow doors and that, so far as he knew, there was no one there to close the doors. He had put the chain across after the last car was loaded. There is no reason why the bow doors should not have been closed as soon as the chain was in position. Mr. Ayling was asked whether there was any reason why he should not have shut the doors. He replied "It has never been part of my duties to close the doors or make sure anybody is there to close the doors". He also said "At that stage it was harbour stations so everybody was going to their stations". He took a narrow view of his duties, and it is most unfortunate that that was his attitude. It is only fair to add that his behaviour after the Herald capsized was exemplary. In the absence of any deck officer, he took the responsibility for organising the rescue efforts, first from the bridge and later in the passenger spaces.

1.3 The questions which arise are: Why was the absence of Mr. Stanley from his harbour station not noticed? And, why was there not a foolproof system which would ensure that the vital task of closing the bow doors was performed irrespective of the potential failure of any one individual? This was not the first occasion on which such a failure had occurred. In October 1983, the assistant bosun of the PRIDE had fallen asleep and had not heard "Harbour Stations" being called, with the result that he neglected to close both the bow and stern doors on the sailing of the vessel from no. 5 berth, Dover.

1.4 A general instruction issued in July 1984 prescribed that it was the duty of the officer loading the main vehicle deck (G deck) to ensure that the bow doors were "secure when leaving port". That instruction had been regularly flouted. It was interpreted as meaning that it was the duty of the loading officer merely to see that someone was at the controls and ready to close the doors. That is not the meaning of the instruction. The instruction is not clearly worded, but, whatever its precise meaning, it was not enforced. If it had been enforced, this disaster would not have occurred. We will revert to these points later.

1.5 Mr. Paul Ronald Morter was the Second Officer of the Herald on 6th March. Mr. Morter went to G deck during the course of loading to relieve the Chief Officer. Despite the arrival of Mr. Morter, the Chief Officer remained on G deck for a time, without explaining why he did so. In due course, the Chief Officer left Mr. Morter in charge of loading. About 10 or 15 minutes before the ship was due to sail, the Chief Officer, who had overheard differences between Mr. Morter and the shore staff, returned and, according to a deposition made by him on the 1st April 1987, suggested that the second officer should go aft and stand by for harbour stations while he completed the loading. That statement does not accord with the recollection of Mr. Morter. The evidence of Mr. Morter is that he did not expect the Chief Officer to return before departure. When there were still 20 to 25 cars to load, Mr. Morter overheard on his radio the Chief Officer giving orders. The two officers did not meet face to face. Mr. Morter assumed that once the Chief Officer had arrived and started issuing orders, he, Mr. Morter, was no longer to exercise the responsibilities of a loading officer. The Court sensed that there was some tension between the Chief Officer and Mr. Morter and that the whole picture had not emerged in the course of their evidence. We quote one short passage from the questions put to Mr. Morter by Mr. Owen and the answers thereto.

Q. Was there ever a set routine for loading this vessel?

A. The cargo duties were shared between the two officers on the ship, not in a set down pattern.

Q. On this occasion, you were in effect relieved of responsibilities of the loading officer within a matter of minutes to go before sailing?

A. Yes.

Q. Was that unusual?

A. Yes.

Q. When the Chief Officer came to G deck and started issuing orders, did you regard yourself as relieved of any responsibility with regard to the closure of the bow doors on G deck?

A. I remained on G deck . . . He took over as the loading officer; so I assumed he took the responsibilities that go with that job.

Q. You say you assumed that; what was your understanding when that happened, did you think you were relieved of all responsibility to ensure that the bow doors were closed?

A. I was not sure, which was why I remained there and discussed it with the Chief Officer before I left the deck.

Q. Discussed the closing of the bow doors with him?

A. No.

Q. Well, what was it you were not sure about, whether you were still loading officer or what?

A. I knew that job had been taken away from me.

Q. What were you not sure about?

A. I discussed with the Chief Officer whether I would go aft, and that was what I was clarifying with the Chief Officer, I was to go down aft.

Mr. Morter told the Court that if he had remained as the loading officer, he would have communicated with the assistant bosun, and he would have waited for a certain period and then chased after him.

1.6 Although the totality of the evidence left the Court with a sense of unease that the whole truth had not emerged, it was in the circumstances set out above that Mr. Leslie Sabel, the Chief Officer, relieved the Second Officer as loading officer of G deck shortly before he instructed the quartermaster to call the crew to harbour stations. Accordingly, it then became the duty of Mr. Sabel to ensure that the bow doors were closed. He does not dispute the fact that this was his duty. But he, too, interpreted the instruction laid down in July 1984 as a duty merely to ensure that the assistant bosun was at the controls. Mr. Sabel had been working with Mr. Stanley during the day of the disaster, and he knew that it was Mr. Stanley's duty to close the doors. Mr. Sabel should have been able to recognise Mr. Stanley.

1.7 The accuracy of some of the evidence given by Mr. Sabel was challenged at the investigation.

For this reason, it is important to bear in mind the physical injuries and shock suffered by Mr. Sabel. When the Herald began to heel to port, Mr. Sabel was in the officers' mess room.

When he realised that something was seriously wrong, he went to the bridge. He had entered the wheelhouse when the Herald capsized. He lost his footing and was thrown violently to the port side. Water flooded into the wheelhouse and over his head. Mr. Sabel suffered injuries which were still causing him pain at the time when he was called as a witness to give oral evidence. The Court has made due allowance for his physical and mental condition.

1.8 The first recorded statement of Mr. Sabel was made in a deposition dated 1st April 1987. On that occasion he said:—"I then checked that there were no passengers in the bow area likely to come to harm, and ensured that there was a man standing by to close the bow doors, I do not remember who he was. Having ascertained everything was in order on the car deck, I went to the bridge, which was my harbour station, assisting the Master". The evidence which Mr. Sabel gave at the investigation was different. He then said that when he left G deck, there was a man approaching, whom he thought was the assistant bosun coming to close the doors and that the man came within about 20 feet of him. His evidence was that there were passengers on the car deck (contrary to his earlier statement) and that he was distracted. If there was a man approaching, we know that the man was not the assistant bosun, who was asleep in his cabin. Who was it? Certainly, it was not a member of the deck crew, all of whom were in other parts of the ship.

1.9 The body of one of the motormen was found on G deck after the Herald had been salved. It seems highly unlikely that any man would have stayed on G deck for about 20 minutes, while the Herald put out to sea with her bow doors open. Therefore, the presence of that body does not support the evidence of Mr. Sabel. As we have already said, Mr. Ayling thought he was the last man to leave G deck. The Court has reached the conclusion that Mr. Sabel's recollection as to what occurred is likely to be at fault. The probability is that he left the area of the bow doors at a time when there was no one on G deck. It is likely that at that time he felt under pressure to go to the bridge, because that was his harbour station, and that he had confidence that Mr. Stanley would arrive on G deck within a few moments. Mr. Sabel has carried out this operation on many occasions. When he was giving evidence, he may have muddled one occasion with another. The precise facts are of no consequence because, on either version, Mr. Sabel failed to carry out his duty to ensure that the bow doors were closed. He was seriously negligent by reason of that failure. Of all the many faults which combined to lead directly or indirectly to this tragic disaster that of Mr. Leslie Sabel was the most immediate. This Court cannot condone such irresponsible conduct. For this reason, his certificate of competency must be suspended.

13.29 KNOWLEDGE CHECK

The following questions will give the reader the opportunity to check their progress and understanding of the previous chapter content prior to a review of learning outcomes.

The science of analysis of case studies follows quite strict guidelines; however, the solutions and opinions may often be quite subjective, and different assessors may often come to quite different conclusions.

Please use your new knowledge from the previous Parts One and Two to analyse and then answer the following questions. I have included my own personal opinion following the questions for comparison. I do not claim this is the definitive correct answer but simply my opinion based upon my own personal career experiences.

Case Study 4 Questions

1. What are the most safety critical issues you have identified in the case study?
2. List the different factors you have identified.
3. Describe the issues you consider as behavioural.
4. Describe the issues you consider as cultural.
5. Describe the issues you consider as professional.
6. Identify the factors you consider as common denominators in the highlighted case studies.
7. Does it appear that crew were always aware of what their specific duties required of them?
8. Who should be responsible for the quality of training in such a company?
9. Who is responsible for the oversight of all company operations?
10. What conclusion do you make or what is an overall impression about the operation?

Case Study 4 Author Opinion (Answers)

1. The bow doors were not closed at the appropriate time.
2. The assistant bosun fell asleep and did not turn up for duty.
3. The Bosun did not ensure that his crew carried out their duties and assumed the tasks were completed.
4. It appears that the company did not follow Standard Operating Procedures (SOPs), which raises questions about basic training philosophy.
5. A lack of SOPs raises questions about individual and corporate training and regulatory adherence to procedures.
6. It appears that Accountable Managers are often absent from and unaware of the practicalities and issues arising from everyday operations.
7. No. It seems like even though there was little evidence of formal SOPs, the accepted behaviour was disturbed by a non-standard removal of some crew from their normal duties causing some confusion.
8. The Head of Training monitored by officers of the Quality Assurance Department.
9. The Accountable Manager as nominated in the Company Operations Manual.
10. It appears very random and undisciplined with poor training and very little oversight from senior management.

13.30 COMMENT

In October 1987, a coroner's inquest jury into the capsizing returned 187 verdicts of unlawful killing. Seven people involved at the company were charged with gross negligence manslaughter, and the operating company, P&O European Ferries (Dover) Ltd, was charged with corporate manslaughter. Even though the company was ordered to be acquitted by the judge since the tragedy could not be specifically attributed to any individual, it later led to changes in the law permitting the Accountable Manager to be held responsible for employee negligence if it could be proved that he knowingly allowed unsafe practices to exist and made no effort to mitigate such acts.

Case Study 5—Drone Airprox Statistics 2019

Consolidated Drone/Balloon/Model/Unknown Object Report Sheet for UKAB Meeting on 13th Feb 2019					
Total	**Risk A**	**Risk B**	**Risk C**	**Risk D**	**Risk E**
10	**2**	**5**	**3**	**0**	**0**
This is not specifically a Case Study but an extract from several accident case study statistics.					

Airprox Number	Date Time (UTC)	Aircraft (Operator)	Object	Location Description Altitude	Airspace (Class)	Pilot/Controller Report Reported Separation Reported Risk	Cause/Risk Statement	ICAO Risk
2018318	14 Dec 18 1009	Legacy 500 (Civ FW)	Drone	5604N 00315W 8 nm NNE Edinburgh 3,000 ft	Edinburgh CTR (D)	**The Legacy Pilot** reports that he was the PIC. Sealed on the right, operating as PM and providing line training to a new Captain They were in receipt of radar vectors from Edinburgh, downwind right- hand for RW24, Having finished the approach briefing, the PIC looked up and saw something black, moving in his peripheral vision on the right. He turned and looked right and clearly saw a quadcopter' like drone. There was no time to take avoiding action The PIC reported the drone to ATC who informed the police, to whom the PIC gave a statement on landing. The Pilot commented that he was surprised and angry at the drone's proximity and stated that a mandatory identification device should be fitted to drones before a multi-million pound engine is destroyed, or worse. **Reported Separation**: Oft V/20 m H **Reported Risk of Collision**: High The Edinburgh controller reports that the Legacy Pilot reported an Airprox whilst overland in the inanity of Burntisland The drone was not observed on radar	**Cause** The drone was being flown above the maximum permitted height of 400 ft such that it was endangering other aircraft at that location. The Board agreed that the incident was therefore best described as the drone was flown into conflict with the Legacy. **Risk**: The Board considered that the Pilot's overall account of the incident portrayed a situation where providence had played a major part in the incident and/or a definite risk of collision had existed.	A

FIGURE 13.4 Extract from 2019 Airprox statistical report on drone accidents.

Airprox Number	Date Time (UTC)	Aircraft (Operator)	Object	Location Description Altitude	Airspace (Class)	Pilot/Controller Report Reported Separation Reported Risk	Cause/Risk Statement	ICAO Risk
2018321	23 Dec 18 1457	A330 (CAT)	Drone	5126N OOOOOW 4 nm SSW London City 4000 ft	London TMA (A)	**The A330 Pilot** reports that a blue drone was sighted off the right-hand-side of the aircraft, about 200 ft below. The aircraft was 1 nm south of an extended 17 nm final approach to LHR RW27R. **Reported Separation:** 200 ft V/NK H The Heathrow controller reports that at approximately 1457 the A330 Pilot reported seeing a blue drone not very big' In size pass 300 ft below. The A330 was passing 4000 ft at the time and was 4 nm SSW of London City Airport.	**Cause:** The drone was being down above the maximum permitted height of 400 ft such that it was endangering other aircraft at that location. The Board agreed that the incident was therefore best described as the drone was flown into conflict with the A330. **Risk:** The Board considered that the Pilot's overall account of the incident portrayed a situation where although safety had been reduced, there had been no risk of collision.	C

FIGURE 13.4 (Continued)

RISK PROBABILITY	RISK SEVERITY				
	Catastrophic A	Hazardous B	Major C	Minor D	Negligible E
Frequent 1	5A	5B	5C	5D	5E
Occasional 2	4A	4B	4C	4D	4E
Remote 3	3A	3B	3C	3D	3E
Improbable 4	2A	2B	2C	2D	2E
Extremely Improbable 5	1A	1B	1C	1D	1E

FIGURE 13.5 This chart shows the standard ICAO Risk Severity/risk probability matrix.

Airprox Number	Date Time (UTC)	Aircraft (Operator)	Object	Location Description Altitude	Airspace (Class)	Pilot/Controller Report Reported Separation Reported Risk	Cause/Risk Statement	ICAO Risk
2016322	26 Dec 18 1420	DHC8 (CAT)	Drone	5133N 00045E 5 nm E Somhend 6000 ft	London TMA (A)	**The DHC8 Pilot** reports inbound to London/city when the crew observed a black drone with 2 red rotors. The crew observed that they were operating above overcast cloud tops. ATC were informed. **Reported Separation:** betow750 m **Reported Risk of Collision:** High **The London controller** reports that the DHC8 Pilot reported a red drone with 2 propellers. The incident was reported to police.	**Cause** The drone was being flown above the maximum permitted height of 400 ft such that it was endangering other aircraft at that location. The Board agreed that the incident was therefore best described as the drone was flown into conflict with the DHC8. **Risk:** The Board considered that the Pilot's overall account of the modem portrayed a situation where safety had been much reduced below the norm to the extent that safety had not been assured.	B
2016323	30 Dec 18 1845 (Night)	EMB175 (CAT)	Unk Obj	5554N 00420W Glasgow 600 ft	Glasgow CTR (0)	**The EMB175 Pilot** reports that on approach to Glasgow airport, when passing about 600 ft he saw an object pass between 3 and 10 ft from the aircraft, at the same level. He couldn't tell was the object was. it was It up in various places and was more horizontally long than it was vertically. **Reported Separation:** Oft V/3–1 Oft H	**Cause** The Board were unable to determine the nature of the object reported and so agreed that the incident was therefore best described as a conflict m Class D. **Risk:** The Board considered that the Pilot's overall account of the modem portrayed a situation where providence had played a major part in the modern and/or a definite risk of collision had existed.	A

FIGURE 13.6 A second extract from 2019 Airprox statistical report on drone accidents.

Airprox Number	Date Time (UTC)	Aircraft (Operator)	Object	Location Description Altitude	Airspace (Class)	Pilot/Controller Report Reported Separation Reported Risk	Cause/Risk Statement	ICAO Risk
2018324	30 Dec 18, 1832 (Night)	A319 (CAT)	Drone	5135N 00036E 3 nmW Southend 4000 ft	London TMA (A)	The A319 Pilot reports that a suspected drone was seen on the right-hand side. It had a blue flashing light and Twinkling' red light. It did not appear to be moving fast relative to the A319. as it would do were it another aircraft, so the crew surmised that it was just their forward speed causing the drone to pass behind them. The p4ot noted that the blue/red light intensity was low compared to standard aircraft lighting. TCAS did not display an intruder. The incident was reported to ATC. **Reported Separation:** Oft V/1 nm H **Reported Risk of Collision:** Low	**Cause** The drone was being flown above the maximum permitted height of 400 ft such that it was endangering other aircraft at that location. The Board agreed that the incident was therefore best described as the A319 Pilot being concerned by the proximity of the drone. **Risk:** The Board considered that the Pilot's overall account of the modem portrayed a situation where although safety had been reduced, there had been no risk of collision.	C

FIGURE 13.6 (Continued)

Airprox Number	Date Time (UTC)	Aircraft (Operator)	Object	Location Description Altitude	Airspace (Class)	Pilot/Controller Report Reported Separation Reported Risk	Cause/Risk Statement	ICAO Risk
2018325	2 Jul 18 1145	EMB135 (Civ Com)	Drone	5132N 00034W 10 nmWNortholt 1800 ft	London CTR (D)	**The EMB135 Pilot** reports that a red drone was spotted from the cabin by the cabin crew on the starboard side of the aircraft when on extended base leg for Northolt RW07. The drone was reported to be higher than the aircraft and turning away in a westerly direction. Cabin crew reported the incident to the Captain after landing. Reported Separation: NK V/150 m H	**Cause** The drone was being flown at the practical limit of VLOS and in a position such that it was endangering other aircraft at that location. The Board agreed that the incident was therefore best described as the drone was flown into conflict with the EMB135. **Risk:** The Board considered that the Pilot's overall account of the incident portrayed a situation where safety had been much reduced below the norm to the extent that safety had not been assured.	B

FIGURE 13.7 A third extract from 2019 Airprox statistical report on drone accidents.

Airprox Number	Date Time (UTC)	Aircraft (Operator)	Object	Location Description Altitude	Airspace (Class)	Pilot/Controller Report Reported Separation Reported Risk	Cause/Risk Statement	ICAO Risk
2019005	12 Jan 19 0945	A320 (CAT)	Drone	5118N 00003E IVOBIGVOR FL080	London TWA (A)	**The A320 Pilot** reports that they were approaching the BIG hold at FL080 when an object was noticed by both Pilots in the distance. As it approached it became clear it was a drone and was possibly being manoeuvred close to them. It passed by the port wing at a distance of about 30–50 meters. It was large in size and black in colour with twin rotors. No avoiding action needed or taken, but both crew agreed that had it been closer. they would probably have had to manoeuvre to avoid. It was reported to LHR approach controller, and the crew requested to move to another hold to avoid re-crossing the drone next time round the hold. They were rerouted to OCK at FL070. Reported Separation: Oft V/30–50 m H Reported Risk of Collision: Low The Heathrow Int South controller reports that the A320 Pilot reported a large black drone passing down his left-hand-side whilst inbound to the BIG hold at FL080.	**Cause:** The drone was being flown above the maximum permitted height of 400 ft such that it was endangering other aircraft at that location. The Board agreed that the incident was therefore best described as the drone was flown into conflict with the A320. **Risk:** The Board disagreed with the A320 Pilot's assessment of risk and considered that separation was such that the incident portrayed a situation where safety had been much reduced below the norm to the extent that safety had not been assured.	B

FIGURE 13.7 (Continued)

Airprox Number	Date Time (UTC)	Aircraft (Operator)	Object	Location Description Altitude	Airspace (Class)	Pilot/Controller Report Reported Separation Reported Risk	Cause/Risk Statement	ICAO Risk
2019006	9 Jan 19 1458	B737 (CAT)	Drone	5049N 00044W Chichester FL190	London FIR (A)	**A B737 Pilot qualified passenger** reports that he saw a large drone pass on the right side which, after a few seconds, appeared to turn away and start descending. The passenger stated that he was certain the object was not a helicopter or other aircraft. **Reported Separation**: -0 ft V/<500 m H **Reported Risk of Collision**: Not reported	**Cause:** The drone was being flown above the maximum permitted height of 400 ft such that it was endangering other aircraft at that location. The Board agreed that the incident was therefore best described as the drone was flown into conflict with the B737. **Risk**: The Board considered that the Pilot's overall account of the incident portrayed a situation where although safety had been reduced, there had been no risk of collision.	C
2019009	13 Jan 19 1050	A320 (CAT)	Unk Obj	5127N 00002E Heathrow 5000 ft	London TMA (A)	**The A320 Pilot** reports that during descent to Heathrow, passing about 5000 ft on heading 300* he saw an unknown object passing on the left-hand- side of the aircraft slightly below. The object was only in sight for a short moment. It was of a •roundish' irregular shape with black/white/orange stripes on the top and approximately 50 cm • 1 m in size. It could even have been a small parachute. As it all happened extremely quickly, and the object was only in sight for a very short moment a better description of the object or the judgement of the closest point was difficult, but he estimated it to be vertical separation 100 ft and lateral separation 100 m. He reported the event to ATC. **Reported Separation:** 100 ft V/100 m H	**Cause:** The Board were unable to determine the nature of the object reported and so agreed that the incident was therefore best described as a conflict in Class D. **Risk:** The Board considered that the Pilot's overall account of the incident portrayed a situation where safety had been much reduced below the norm to the extent that safety had not been assured.	B

FIGURE 13.8 A fourth extract from 2019 Airprox statistical report on drone accidents.

Airprox Number	Date Time (UTC)	Aircraft (Operator)	Object	Location Description Altitude	Airspace (Class)	Pilot/Controller Report Reported Separation Reported Risk	Cause/Risk Statement	ICAO Risk
2019012	22 Jan 19 1026	C406 (Civ FW)	Drone	5206N 00100W Silverstone FL067	Daventry CTA (A)	**The C406 Pilot** reports flying survey lines when he saw a large dark coloured drone pass by on the left-hand side. **Reported Separation:** Oft V/50 m H **Reported Risk of Collision:** Low	**Cause:** The drone was being flown above the maximum permitted height of 400 ft such that it was endangering other aircraft at that location. The Board agreed that the incident was therefore best described as the drone was flown into conflict with the C406. **Risk:** The Board considered that the Pilot's overall account of the incident portrayed a situation where safety had been much reduced below the norm to the extent that safety had not been assured.	B

FIGURE 13.7 (Continued)

Case Study 5 Drone-Specific Questions

In this section showing statistics from the Airprox 2019, the task is to just read the details and analyse any trends specific to drone operations.

1. After analysis of the Drone Airprox statistics, can you observe any common trends?
2. List those trends.
3. How would you mitigate such trends escalating in the future?
4. What other common dominators can you observe?
5. Describe what regulatory changes you consider appropriate?

Case Study 5 Author Opinion (Answers)

1. On most of the reports, the drone was flying above the legal maximum height of 400 feet.
2. According to the ICAO RISK SEVERITY/PROBABILITY CHART, the instances were frequent and hazardous.
3. Better training, more restrictive regulation, more oversight of operators.
4. Most incidents were near large airports.
5. Confiscation of equipment and stringently enforced fines for serial offenders.

13.30.1 Comment

Since drone operations are increasing rapidly, there is obviously more chance of incidents if people do not have the appropriate understanding and training. As they increase in size, complexity, and mission capability, much more regulation will be required, and ultimately the same restrictions will become more compatible with manned aviation.

13.31 LEARNING OUTCOME REVIEW

LO 1: Analyse each case study to investigate what factors you think have the biggest influence in causing the occurrence.

LO 2: Identify any common structural or behavioural factors influencing the occurrences.

LO 3: Describe the importance of having appropriate structures to facilitate the identification and classification of outcomes.

LO 4: Identify and explain which factors you feel are the most challenging to isolate.

Part Four

Education

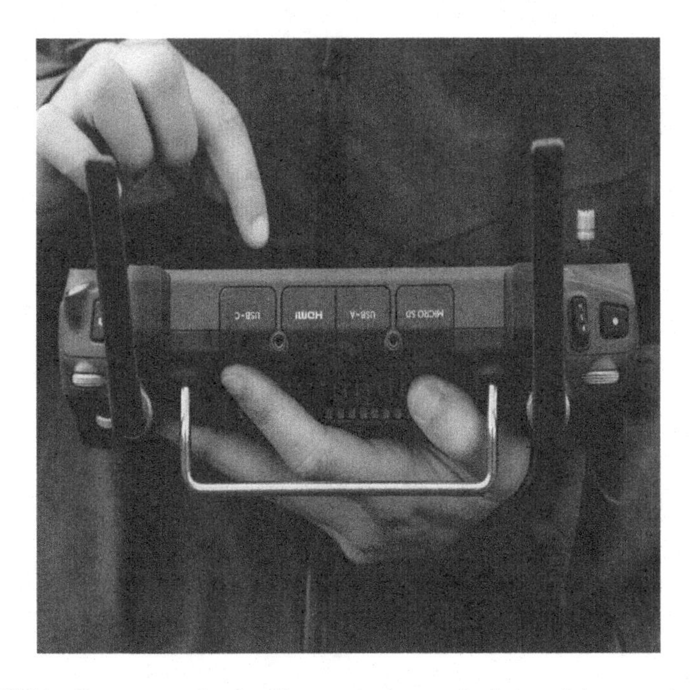

FIGURE PIV.1 Drone control unit with an operative manipulation of the controls.

DOI: 10.1201/9781032620220-17

14 Teaching and Learning

14.1 LEARNING OUTCOMES

By the end of the lesson, students will be able to:

LO 1: Analyse which factors are essential for all training programmes.
LO 2: Investigate appropriate processes to understand, identify, and mitigate those behaviours which affect an individual's ability to learn.
LO 3: Identify and explain why we all possess different learning styles.
LO 4: Investigate methods as to how we can accommodate these different learning styles.

14.2 PRIME OBJECTIVES FOR AN INSTRUCTOR

It is essential that all instructors understand the importance of determining the level of class members' knowledge and experience and an outline of a student's background prior to delivering any training course.

This will assist with determining the academic level of complexity that students will find comfortable when learning.

If at all practical, this can be best achieved by asking everyone to introduce themselves to other class members with a short history of their experience and asking what they hope to gain from the current training session. This should give the instructor a basic overview of the class members' personalities, attitudes, and experiences as well as break the ice to help with developing a group/team spirit.

Essentially, an instructor should gauge a student's baseline level of knowledge, skill, and attitude. Then using appropriate methods, they should help them develop and acquire all relevant areas safely, competently, and comfortably.

14.3 HOW CAN THIS BE ACCOMPLISHED?

Effective instructors must be flexible and understanding enough to realise that the learning journey can be radically different for each student. They must be able to relate to any issues the student may have in understanding a lesson or concept. This understanding can often be easier if they can clearly and honestly understand their own behaviour and how that can affect others.

DOI: 10.1201/9781032620220-18

14.4 ESSENTIAL TRAINING REQUIREMENTS

A good instructor is able to develop quickly in some key areas. By making an effort to get to know their student, it will allow the instructor to make better decisions on how to tailor the approach to different teaching styles. The areas that an instructor should be able to develop quickly are explained in the next sections.

14.5 TRUST

Building up a bond of trust with the student can sometimes take time, and it requires a consistent track record of showing interest and care in their development. However, once established, it can vastly improve student performance by helping the instructor share important concepts as they appear more acceptable.

14.6 FLEXIBILITY

A good instructor must develop the ability to make quick and accurate assessments of how a particular concept is being understood and accepted by the class.

They will often need to be flexible and adaptable to completely alter their teaching materials and concept to a way that the class may understand easier. This can be a difficult challenge, and highlights the need for the instructor to have an in-depth subject knowledge so they can approach the subject from many different perspectives. Often, a student's performance can vary drastically from day to day. They must be aware of influencing factors and be able to make modifications to accommodate this.

14.7 TIME MANAGEMENT

This can be one of the most challenging areas to develop. It should improve with experience and can be facilitated by adequate preparation and planning. It is so easy to lose track of time if diverted by events such as in-depth discussion or a high volume of questions from students. Scientific research (the Technique for Human Error Rate Prediction, THERP) has shown that poor time management increased the escalation factor of making human errors by a factor of 11!

14.8 DIPLOMACY

In a commercial or industrial context, an instructor often finds that they have to walk a fine line between advocating what their training has taught them and quite often, conflicting commercial, and organisation pressure so it is necessary to carefully consider what you are trying to achieve.

14.9 ORGANISATIONAL POLICIES

An effective solution when involved in such conflicts of interest is to deliver both answers with the caveat that company standard procedures would dictate a specific approach. This can be followed up with the advice to open a discussion with the relevant responsible person, with a suggestion for considering possible changes, along with suggestions, and documentary evidence as to how the company may benefit from such change.

14.10 INTERVENTION

At what part of a student's learning journey is it appropriate to intervene and correct them? Since we are all different, there is no hard and fast approach to this. Therefore, it is important to gauge the individual's attitudes and reactions and then tailor the approach accordingly. If a student is pushed too hard when close to achieving the objective, it can be demotivating. Alternatively, others may just lack the confidence of giving an answer for the fear of being incorrect and embarrassment in front of their peers. This is a delicate balance but can be best estimated by having a good understanding of the student.

14.11 THE STUDENT

Another major factor would depend on whether the situation is practical or theoretical, for example in a flight training, an instructor may have to take the controls if a situation is becoming too dangerous without having time to explain their actions, compared to theoretical training where the instructor has the luxury to wait longer to let the situation develop, in the hope the student sees the error.

14.12 INSTRUCTOR MOTIVATION

It is essential that an instructor can motivate the students to achieve the highest standards. One of the best motivations an instructor can provide is observing the student journey from an unknown to a known and the obvious awakening of a new subject interest that they have helped develop.

14.13 STUDENT MOTIVATION

An instructor must possess a good subject knowledge and display a positive optimistic "can do" attitude. If they are passionate about their subject knowledge and display natural enthusiasm, it will very likely rub off on the students. It is important to establish why are the students attending the class. Have they been instructed to attend, or do they really want to attend? Their motivation, or lack of, could significantly impact upon the structure and approach of an instructor.

14.14 DIFFERENT LEARNING STYLES

When designing a training session, it is helpful to understand what different learning styles and combination of styles represent the target audiences' abilities and aspirations. It has been suggested that for a good training response, a concept should be presented in as many forms as possible to ensure that at least one method is ideal

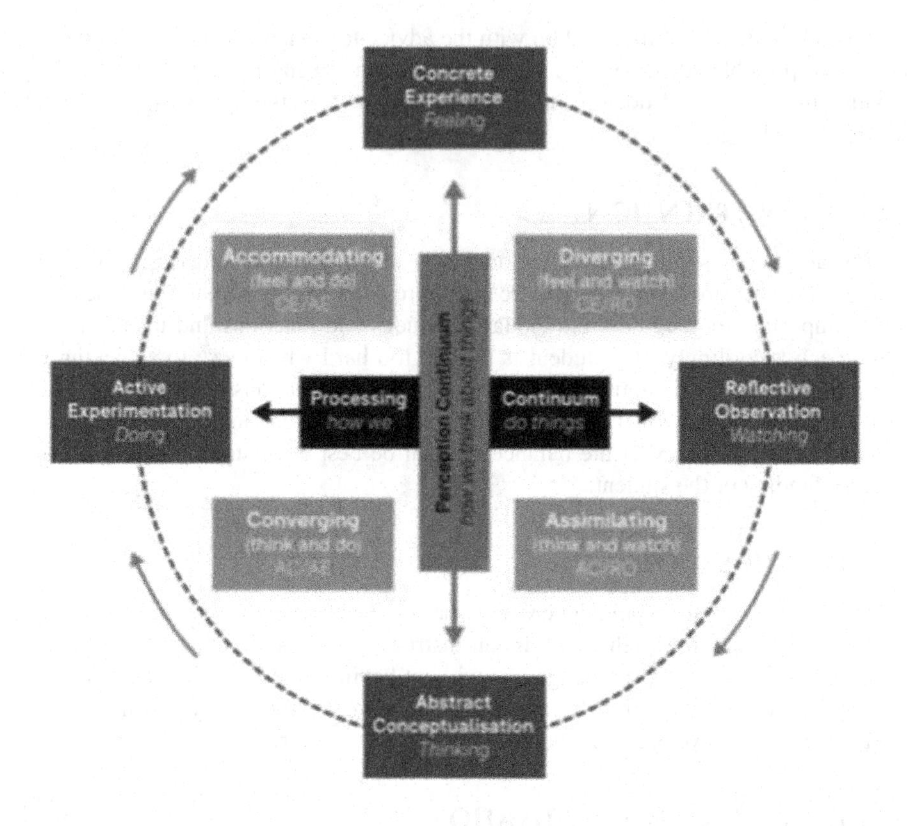

FIGURE 14.1 Kolb's Learning Cycle.

for everyone to respond and learn from. The diagram given in Figure 14.1 illustrates "Kolb's Learning Cycle".

The four outer areas looks at the stages of learning;

- Concrete Experience: Doing or having an experience
- Reflective Observation: Reviewing or reflecting on the experience
- Abstract Conceptualisation: Learning from the experience
- Active Experimentation: Planning or trying out what has been learnt.

The four inner areas, seen here in, looks at the different learning styles as described in the following sections.

14.15 DIVERGING

This style takes a creative approach. Individuals who prefer this style often value feelings and take an interest in others. The Diverging learning style suits hands-on activities or the standard class lecture.

14.16 ASSIMILATING

This style emphasises reasoning. Individuals who are drawn to this style are usually able to analyse the experience in its entirety and are able to review all areas and facts. The Assimilating learning style suits project designing and is best approached through pre-prepared exercises that the student can complete with minimal instructor input.

14.17 CONVERGING

This learning style highlights problem solving as an approach to learning. Individuals who prefer this style tend to be able to make decisions and apply their ideas to new experiences. Some instructional techniques that this style suits are workbooks or worksheets, computer-based tasks, or interactive activities.

14.18 ACCOMMODATING

This style is adaptable and intuitive. These individuals usually prefer trial and error, preferring to discover the answers themselves. They usually have good people skills. Some instructional techniques that this style prefers are activities that allow them to be engaged, exploration with instructor support for questions.

14.19 HOW CAN A TRAINING PROGRAMME BE DESIGNED TO MAXIMISE EFFECTIVITY?

All effective training requires a structured approach. To ensure this, can we use the SMARTER acronym mentioned earlier in this handbook.

14.20 SPECIFIC

It is essential that with any training course, the Learning Outcomes presented to students are clearly defined. It has been demonstrated that when students are given a specific set of criteria they remember as much as 7% more of the information delivered to them. Performance is up to 7%.

14.21 MEASURABLE

All goals must be measurable from the perspective of how to know if targets are being achieved, and allowing specific quality assurance measurement.

14.22 ACHIEVABLE

As part of the standardisation process, all training courses must move towards being achievable and realistic in that they lead the student to reaching a conclusion or standard of competence which is relevant to what they are trying to achieve.

14.23 RELEVANT

To engage the student in a course while maintaining and stimulating their interest, the content must relate to topics that are important and relevant to the goal they are working to achieve.

14.24 TIMEBOUND

A timebound goal is essential for pacing a lesson, ensuring that a goal is met within a specified timeline. A good example is illustrated in the Safety Management Systems Risk Mitigation process when a corrective action must be completed by a specific date for practical and regulatory reasons.

14.25 EVALUATED

A process must be evaluated in order to monitor Quality Assurance and collect information. This forms the basis of any modifications which may be required to the Training Programme. The statistical data may also be collated to form part of the auditing process. Internal organisation or regulatory requirement. This data will also be required for the next step in the Safety Cycle, the review.

14.26 REVIEWED

Continuous review is an integral part of any training programme to ensure standards are being maintained and to introduce any new concepts or developments in that subject matter. A documented paper trail is not only a regulatory requirement in accredited training, but it also facilitates the integrity of the overall course content.

14.27 SKILL FADE

With all training programmes, it is worth considering that skills will often deteriorate without regular practice. If for example a Pilot or a surgeon does not practically perform their core duties on a regular basis, it is generally accepted that they should participate in some appropriate refresher training. In Commercial Aviation, there is a mandatory recency requirement of performing 3 take offs and landings within every 90-day period.

14.28 TRAINING FACILITATION

14.28.1 Overview

In the beginning of this chapter, we looked at some of the basic principles of teaching and learning. In this section, we focus upon specific practical techniques to make that task easier and more efficient. The actual mechanisms of content delivery and assessment are critical and so important for highlighting how to manage these processes to maximum benefit.

14.29 WHAT IS FACILITATION?

Oxford English Dictionary definition:

> The action of facilitating something—"third party facilitation seeks to promote the resolution of conflict".

Facilitate means making (an action or process) easy or easier. "Schools were located in the same campus to facilitate the sharing of resources". Origin (see facile): In early 17th century from French word *faciliter*, from Italian word *facilitare*, from facile—easy. Facilitation is a learnt skill which describes a method of making a process such as negotiation, conflict management, and training de briefing flow much easier and smoothly, in a structured and orderly way. It places emphasis upon the students sharing their knowledge and experiences to come to their own agreed conclusions, rather than being told what the conclusions should be. Educational research demonstrates that lessons are absorbed and remembered better, when the student comes to the conclusion about a concept themselves, rather than being spoon-fed many facts. Facilitation is one of the most basic rules of training any subject. Where the student needs to be taken from an unknown to a known, this journey is best accomplished by using methods appropriate to the abilities and learning styles of the student. To establish what is an appropriate method, the instructor must first get to know the student, understand their ability and potential, and then modify the subject delivery at a pace and complexity to suit the student. The art of facilitation is the ability to be flexible enough to understand what competency you are dealing with and how to maximise the student understanding.

14.30 COURSE INTRODUCTIONS

Here, we have four key areas that an instructor should remember when delivering a course introduction:

- Do not forget the safety briefing
- Clarify the role of instructor and detail the expectations for crew participation
- Tell the students how long the session will last
- Do not cut sessions short for high-performing crew

There are several examples of where an instructor has omitted a Safety Brief at the beginning of the learning session, and then an emergency such as a fire was experienced. The subsequent injuries and even deaths were largely attributed to the instructor's failure to adequately promote the required health and safety regulations, and they were considered to be legally responsible. It only takes a few minutes, so cover yourself by performing the adequate safety briefs.

14.31 THE BASICS

The following list of points represents some of the basic components of the process of how facilitation may be organised:

- Keep any discussions student-centred
- Encourage students to participate actively
- Adapt your level of input using the capabilities of each student
- Balance your dual role as an instructor and facilitator
- Reinforce good student performance
- Show you are interested in what the students have to say with your attitude
- Try not to be condescending or make long speeches
- Do not interrupt or move on from a topic while a student still wants to discuss it
- Use questions to promote in-depth student participation
- Follow up on student topics and redirect student questions and comments back to them
- Ask questions with pronouns what, who, and why to encourage deeper discussion
- Ensure that all students are drawn into any discussions immediately responding to your question
- Reword questions rather than giving answers
- Use active listening to encourage a continued participation
- Encourage crew members to address each other directly
- Get the crew to talk about what went well
- Get the crew to talk about what could be improved and how
- Push the crew to go beyond just describing what happened
- Ask follow-up questions that require in-depth analysis
- Ask the crew to analyse why they made the decisions they made
- Encourage the crew to discuss the factors that enabled their success

A successful facilitation process promotes deep self and group analysis by the students. The aim is to conclude what went well, and what did not go so well. Any follow-on discussions should suggest what could be improved and how that could be achieved.

14.32 FEEDBACK AND REVIEW

In Chapter 8 Section Communication, the "Communication Loop" demonstrated that the feedback component was integral to obtaining a complete understanding of what message was being delivered. Exactly the same principle applies to all operations where a message or concept is being shared. Objective feedback is very often never delivered, leaving the communicative process misunderstood.

14.33 DELIVERING FEEDBACK

Feedback is probably one of the most badly handled aspects within management and the easiest to perform correctly.

There are two types of basic workplace feedback. The first being Reinforcing feedback. This is where we are confirming that the task or operation is being handled correctly or has been completed to a satisfactory standard.

The second type of basic feedback is Redirecting. This is where guidance is given on the current task progress, while suggesting improved methods to achieve the goal.

Feedback must be based on fact and not opinion. It must be clear, genuine, and unambiguous. It must also be constructive, developmental, and supportive.

Benefits of continuous feedback include:

- Enhances performance
- Boosts motivation
- Improves engagement
- Promotes self-awareness

The typical feedback process includes:

- Preparation
- Arranging a meeting
- Sharing an empathetic message
- Describing the behaviour and its impact
- Starting a dialogue
- Offering suggestions
- Agreeing on the next steps
- Saying "Thank You"

14.34 DELIVERING A "WELL-RECEIVED MESSAGE"

Delivering a well-received message is the ideal goal of objective feedback and can be an inspirational opportunity to motivate your team. If you follow the facilitation techniques covered in this book, they will allow you to take the maximum advantage from your feedback summary.

- Be specific and direct
- Focus on behaviour not character
- Stick to the facts

Feedback has one primary purpose to enhance employee performance: Think empathetically and consider what employees need to hear so they can continue to improve.

14.35 MOTIVATING YOUR TEAM

Most employees wish to do a good job and be recognised for their achievements. A good manager should motivate their team to not only improve the productivity, but also it will make their own tasks that much easier.

A good manager will also understand their team and be able to focus on attitudes and behaviours which can often indicate the basic drivers which motivate individuals.

Two common types of motivation are described in the next sections.

14.36 INTRINSIC

An inner drive that propels a person to pursue an activity, not for external rewards, but because the action itself is enjoyable—a person is motivated by fun, challenge, or satisfaction involved with an activity, not for an outside outcome, pressure, or reward.

14.37 EXTRINSIC

This refers to behaviour that is driven by external rewards, such as money, fame, grades, and praise.

All humans are wired slightly differently and may not react as expected in certain situations. Therefore, an effective manager must be tuned into these differences in attitude, ability, and behaviours.

They must be flexible and skilled enough in understanding, managing, and utilising these differences to maximise individual and group potential and performance.

The best managers will find out what drives their employees and then apply customised strategies to each team member. This may be achieved by:

- Checking your assumptions or pre-conceived ideas
- Asking direct questions
- Observing and analysing behaviours
- Keep talking but mainly watching and listening

Remember, every employee is unique, and not all people are motivated by the same things. Motivation is an individual experience.

14.38 KNOWLEDGE CHECK

The following questions will give the reader the opportunity to check their progress and understanding of the previous chapter content prior to a review of learning outcomes.

1. Why is education such an important component of all organisations?
2. What are the different stages of training which need to be delivered to all organisation members?
3. What is meant by Skill Fade?
4. List five examples of factors required to create a good learning environment.
5. Explain the term different learning styles.
6. List five examples of how to motivate your students.
7. What is the importance of having an objective feedback?
8. Describe two methods of delivering feedback.
9. What is meant by facilitation?
10. How can you facilitate a debriefing session?

14.39 LEARNING OUTCOME REVIEW

Before continuing to the next chapter, it is worth reviewing whether or not you believe you have achieved and fully understood the Learning Outcomes from Part Four.

LO 1: Analyse what factors are essential for all training programmes.
LO 2: Investigate appropriate processes to understand, identify, and mitigate those behaviours which affect an individual's ability to learn.
LO 3: Identify and explain why we all possess different learning styles.
LO 4: Investigate methods as to how we can accommodate these different learning styles.

Part Five

Leadership and Management

FIGURE PV.1 DJI docking station.

DOI: 10.1201/9781032620220-19

15 Leadership and Management

15.1 OVERVIEW

In this module, we will be evaluating the general principles of effective management while pulling together the other elements covered in this course and investigating how to best apply these principles in everyday practice. Safety Management Systems describe a tried and tested structure which can be adopted by any organisation. The Human Factors give us an understanding of how and why we all behave as we do in groups and as individuals. Teaching and learning give us insights into how different people lean in different ways and some guidance on how to make that process more effective. Facilitation further investigates methods by which it is possible to reinforce your teaching techniques to improve and understanding and performance. All these factors together give us a theoretical insight into systems and structures that are needed to be used to become a successful manager. However, by far the most important piece of the puzzle is the bit that is needed to complete the picture. You. As a manager, you apply your own attitude and behaviour to learn and to understand people, what drives them, and how to encourage them to perform to their maximum potential.

15.2 LEARNING OUTCOMES

By the end of the lesson, students will be able to:

LO 1: Understand the management needs for an Unmanned Aircraft Systems Operations Manager.
LO 2: Evaluate theories of management.
LO 3: Analyse how the practical application may affect own organisation's operations.
LO 4: Describe which management style you feel represents your own style.

15.3 YOUR ROLE AS A MANAGER

A manager and a leader are very different roles within an organisation. A manager is someone within an organisation who is responsible for controlling, planning, and

DOI: 10.1201/9781032620220-20

organising. These three factors are structural and largely decided upon by the company hierarchy or rank structure.

Whereas a leader is much more dependent on the abilities of an individual.

Good leadership is a quality which often determines the effectivity of a good manager, but they are not necessarily mutually exclusive. Applying the structures of good management can be learnt but without understanding what drives people and how to motivate them, it may not be possible to become a good leader.

A good leader must be able to understand and relate to the feelings of their team, even if their subsequent actions and decisions do not necessarily reflect this understanding.

What is effective management and leadership?

To be an effective manager or leader, it is necessary to formulate an action plan, taking into consideration the appropriate targets, with timelines factored for costs and expenses. Adequate time for update and review should be considered.

It is also important to consider the respective needs and requirements of a manager and a team, in terms of practicality and reality checks.

It is important to establish a working culture to empower individuals to make objective contributions and express reservations with confidence, promoting team unity.

15.4 TASK COMPLETION

Tasks must be clearly defined and given a projected timeline to completion. This timeline must be achievable, and regular reviews with updates will be required.

15.5 COST

The cost of any projects must be clearly defined and regularly reviewed with a sensible contingency fund included.

15.6 TEAM UNITY

Teams must operate with clear communication and regular supervision to ensure that they are working productively and harmoniously.

15.7 MANAGER AND LEADER NEEDS

The manager must clearly specify their requirements and ensure that they are being actioned.

Team member needs managers should ensure that all tasks allocated are within the capacity of the team members. The team members must feel comfortable to routinely ask questions if they are uncertain of their obligations.

15.8 WHAT MAKES A GOOD MANAGER?

To be successful as a manager requires a comprehensive blend of competencies. They must have sufficient knowledge in their specialised field; but it is equally important for a manager to possess good general knowledge to allow them to relate to their team member needs. Ideally, they should possess a good deal of relevant, high-quality experience.

A manager's individual style of management will determine how their team members react to them, and this will impact the overall individual and team performance.

Next, we detail some important competencies for a successful manager.

15.9 KNOWLEDGE

A good manager must have excellent subject knowledge as well as general knowledge to be able to create solutions and inspire confidence in their team.

15.10 EXPERIENCE

This is one of the most important aspects of good management. However, it must be experience relevant to their role.

15.11 STYLE

Ideally, a manager still should be standardised and disciplined but, at the same time, empathetic and understanding to the needs of the team.

15.12 ENVIRONMENT

It is vital to create a safe, productive, and happy work environment in order to maximise the productivity of the team.

15.13 AMBITION

An ambition to succeed is vital for an effective manager, in order to continually strive for improvement and achieve personal and financial targets.

15.14 WHAT ARE THE QUALITIES OF GOOD MANAGEMENT?

If a manager is to lead by example and project positive influences towards their team, it is obvious the success will largely depend upon the image or persona that they present.

Good communication skills are essential for an effective manager, and actively listening is probably the most important aspect of the role. Only with good listening skills can a manager truly engage with their team and develop the relationships necessary for success.

Leading by example will inspire the team, so if a manager displays confidence and passion about what they do, these characteristics explained in next subsections will be adopted by the team.

15.15 SOME IMPORTANT QUALITIES OF GOOD MANAGEMENT

15.15.1 Confidence

In order to inspire confidence in their team, a manager must behave in a confident manner to help the team develop their own self-confidence.

15.15.2 Inspiring Others

Creating an inspirational group dynamic allows your team members to develop in a much more productive way. This is best achieved by leading by example.

15.15.3 Commitment and Passion

Displaying positive behaviours, with an abundance of commitment and passion, will rub off on team members.

15.15.4 Communication Skills

Good communication skills are necessary to motivate and empower a team. Listening carefully to their comments is probably the most important aspect of good communication.

15.15.5 Decision-Making

A manager must be decisive but flexible enough to admit when they are wrong. They must be positive but not afraid to review their actions and modify the team's instructions if needed.

15.15.6 Accountability

A good leader must be accountable for their decisions. They must demonstrate a strict adherence to Standard Operating Procedures.

15.15.7 Delegation and Improvement

Improvement can only follow continuous monitoring, review, update, and training. A good manager delegates as many duties as they feel confident to allow them to manage their time more effectively.

15.15.8 Creativity and Innovation

These are the building blocks of a sustainable business to prevent stagnation. It also promotes expansive and innovative thinking.

15.15.9 Empathy

A good leader is empathetic with their team. They try to understand how they think and feel.

15.16 NECESSARY MANAGEMENT SKILLS

15.16.1 Adopting a Management Persona

The performance of a team relies heavily upon the working atmosphere the manager promotes by displaying a positive management persona. Ideally, the manager wants the team to follow their lead, remembering that their own behaviour will always be in the spotlight.

A new managers' success will partly depend upon how they project their management persona. Remember to lead by example, your team will look up to you for an example on how to behave. Take into account the best qualities of other managers within your organisation or who has managed you previously. Try to mirror their best qualities.

Try to avoid drastic policy changes, as this will create an unstable atmosphere in the workplace. Although your team may be your friends, or have previously been your peers, you must remember that you are a boss and not a buddy.

- Focus on fairness.
- Have performance-related discussions.
- Avoid favouritism.
- Make yourself visible.
- Be accessible to your team.
- Do not hide behind emails.
- Be a role model.
- Be consistent.

15.17 MANAGEMENT MISTAKES

Management can be a minefield of potential mistakes for new and experienced managers. We will look at some common mistakes to hopefully help.

15.18 MICROMANAGING AND DELEGATING

One of the most common mistakes a manager can make is either micromanaging or over delegating. A manager who tries to micromanage will become overloaded with their own workload. Over delegating will risk losing the respect of the team.

15.19 SETTING VAGUE EXPECTATION

A team or an individual without a clear goal risk becoming unmotivated and their performance will be difficult to gauge.

15.20 OVERCOMMITTING TEAM MEMBERS

This means members becoming overworked and will result in a significant drop in performance and morale. Giving team members too many tasks or not allowing "downtime" risks the team or team members becoming overworked and will result in a significant drop in performance and morale.

15.21 ADOPTING A ONE-SIZE-FITS ALL APPROACH

When it comes to motivating team members, managers need to remember that not all members will be motivated by the usual cash incentives. Each member will be motivated differently, from areas such as responsibility, promotions, work–life balance, to the standard cash motivations.

15.22 UNDER-COMMUNICATING

When it comes to planning and creating strategies for the organisation and team, it is important to acknowledge the ideas of the team. Your team will be the ones who are dealing with the day-to-day running of the operations and tasks, so to disregard their ideas would be a mistake.

15.23 BECOMING A UAS OPERATIONS MANAGER

Becoming a UAS Operations Manager for the first time can be a daunting experience! Managers will be unsure if they will be successful, or wondering what others think about them and how others would judge them. There is no magic formula into success, it is all common sense. Be yourself, natural, analytical and get to know your team while letting them get to know you. Some strategies for a smooth transition to a new managerial role include:

- Reintroduce yourself as a manager.
- Have honest conversations.
- Avoid sweeping changes initially.
- Stay humble.
- Let your actions speak for themselves.

Remember, you were promoted to a manager for a reason. Lead with confidence and take ownership of your new role.

15.24 MANAGEMENT STYLES

Management styles are the different ways that a manager can lead their team. All managers are unique, but there are certain traits that all good managers share.

Being able to recognise your own management style will enable you to develop your skills and become a better leader.

In the next sections, we will explore some of the more common managerial styles.

15.25 LAISSEZ FAIRE

Laissez Faire from the French translates as "Let it go" or "Let it be".

A Laissez Faire management style is when a manager has an attitude of trust and reliance on their team. Laissez Faire managers do not tend to get heavily involved or try to micromanage their team; instead they let their employees use their own creativity, resources, and experience to meet their goals.

Some key traits of Laissez Faire management are:

- There is not much guidance from the manager.
- Employees are expected to solve their own problems.

- They provide constructive criticism when it is required.
- Will take charge when it is required.

15.26 AUTOCRATIC

This management style is where all the decision-making and power are in a single person with very little input from others.

An autocratic manager is the difference between having the attitude, "What should we do?" and

"You do this".

It is a very common management style, it has its clear downsides, but can be very successful.

Steve Jobs from Apple was a famous autocratic leader. He was famous for shouting down ideas he did not like and changing project deliverables during the final stages. He was also famous for firing employees who he felt did not meet his standards.

15.27 KEY TRAITS OF AN AUTOCRATIC MANAGEMENT STYLE

- Managers rarely accept advice from others.
- Micro-manage all decisions, especially important decisions.
- They create lots of rules and thrive in a rule-driven atmosphere.
- They take control over working methods and processes.

15.28 DEMOCRATIC

This management style happens when the group participates in decision-making processes. Everyone is given an opportunity to get involved and share their own ideas.

A democratic leader ultimately decides what decisions are made but focuses on group equality and promotes employees putting forward their own ideas.

Democratic leadership styles are another common style, and they promote morale and productivity.

15.29 KEY TRAITS OF A DEMOCRATIC MANAGEMENT STYLE

- Managers have lots of ideas due to the group discussions.
- Employees feel more committed to the workplace.
- There is an increase in productivity due to employees feeling involved.
- There is a risk that introverted individuals could have their ideas ignored.

Google is an example of a democratic-style workplace.

15.30 TRANSACTIONAL

This management style happens when the manager relies upon rewards and punishment to promote the performance of their employees. This style assumes that employees lack motivation and require supervision in order to complete tasks.

This style relies on employee's self-interest to accomplish tasks.

15.31 KEY TRAITS OF A TRANSACTIONAL MANAGEMENT STYLE

- It is good for achieving short-term goals quickly.
- Tends to get consistent results.
- Employees have clear responsibilities.
- It can stifle creativity.
- It lacks innovation.

Bill Gates from Microsoft is an example of a leader having transactional management style.

15.32 TRANSFORMATIONAL

This management style is passionate and energetic. It focuses on nurturing all employees to succeed.

Transformational leaders encourage creativity and encourage employees to develop new methods of working. This style provides employees with opportunities to develop themselves.

15.33 KEY TRAITS OF A TRANSFORMATIONAL MANAGER STYLE

- It is good for stimulating employees intellectually.
- It fosters a supportive environment.
- Leaders of this style tend to be inspirational.
- Leaders of this style are usually a role model to employees.

Barrack Obama and Jeff Bezos are examples of leaders having a transformation leadership style.

15.34 SETTING TEAM GOALS

Effective goals follow a process.

While managers ultimately may have the final say, the goal-setting process is best viewed as a collaborative dialogue.

All goals should be SMARTER.

- Specific
- Measurable
- Achievable
- Relevant
- Timebound
- Evaluated
- Reviewed

We will be expanding on the SMARTER goal setting later in this publication.

15.35 TEAM GOALS

Effective goals follow a process. When setting goals, it is most effective to clearly define and explain the process to your team.

15.36 REFERENCE COMPANY GOALS

What are you trying to achieve? Why, how, when, and by who?

15.37 CREATE RELEVANT TEAM GOALS

When creating team goals, assign tasks to the respective subject matter expert (SME) in your team.

15.38 ESTABLISH INDIVIDUAL GOALS

Break down the overall task into smaller, manageable goals and delegate them appropriately.

15.39 SET SMALLER MILESTONES

When setting a task, try to establish smaller components which can be achieved over a more even timeline.

15.40 CREATE AN ACTION PLAN

Create and share your plan to monitor the task progress. Ensure you review the team's achievements and update the plan as appropriate.

15.41 HOLD YOUR TEAM ACCOUNTABLE

Setting team goals is only half the equation. To achieve success, you must also help employees follow through with these goals by tracking progress and scheduling reviews.

15.42 COMMUNICATING EFFECTIVELY

Once again, we note the importance of a good effective communication as a prerequisite for nearly all processes.

The fundamentals to effective communication are:

- Using clear and concise language
- Actively listening
- Asking thoughtful questions

In order to assist with being successful, some considerations include:

- Setting clear expectations
- Being direct and straightforward

- Communicating openly and honestly
- Using multiple channels
- Being approachable
- Schedule regular check-ins
- Embrace disagreement

15.43 TEAM RELATIONSHIPS

15.43.1 Building Relationships with a Team

For a manager to build and maintain a good relationship with their team not only improves performance, but it also makes their job a lot easier. It can take a long time to build up mutual trust and respect, but it can be lost in an instant.

Often, once that bond has been broken, it can be impossible to rebuild.

Building professional relationships requires bringing value to the team members and showing them that the manager is invested in their performance. Foundations of a strong working relationship include:

- Mutual respect
- Mutual trust
- Shared understanding
- Sense of goodwill or camaraderie

To develop positive relationships with your team, you need to show members that you care.

This can be achieved by understanding their emotions and offering appropriate support. Understanding what drives and motivates your team, such as giving praise when deserved but at the same time giving constructive criticism with advice, can reinforce their trust and respect.

In order to build a positive working relationship, it requires a combination of the following:

- Understanding team member's emotions
- Offering support
- Giving recognition and praise
- Expressing positivity and enthusiasm
- Showing care for the team

15.44 MANAGING EMPLOYEE PERFORMANCE

The goal of performance management is to keep employees informed.

Everyone should know what they are doing and how they are contributing, as well as how they can continue to learn, develop, and improve.

It should be an ongoing year-round process, not just annual or bi-annual.

Once again, the principles of the Safety Cycle can be referenced—continuous monitoring, review, update, and training. In practice, this can be achieved by having regular one-to-one discussions with your team in addition to group discussions.

15.45 DELEGATING TASKS

Most successful managers develop the art of delegation. Just because they are the designated manager, it does not mean they are the most capable or talented member of the team.

With humility and good communication skills, it is quite simple to identify which members of the team possess which competencies.

The manager is not expected to be able to do all of the work all of the time. They must identify, delegate, and trust in the abilities of the nominated subject matter experts (SMEs). This not only frees their own individual capabilities but also improves job satisfaction.

Delegation is a divide-and-conquer approach. It requires splitting up a project and assigning each person their own individual piece. By delegating tasks, they encourage the entire team. It is essential that when delegating, consideration is given to what is being delegated and to whom. When delegating tasks, the following are required:

- Communicating expectations
- Providing training and support
- Giving an appropriate level of authority
- Giving feedback and celebrating success

15.46 LEADING EFFECTIVE TEAM MEETINGS

The purpose of a team meeting is to outline the goals, delegate tasks, determine sensible timelines, and review actions as per the Safety Cycle.

An effective team meeting needs to be structured and planned and should follow the process we have looked at with the modules covered during this course.

Organising an initial successful meeting should include:

- Defining the objective
- Identifying the participants
- Developing and sharing an agenda
- Assigning any pre-work
- Post-meeting review of any takeaway pointers

15.47 COACHING EMPLOYEES

Learning is a dynamic process and needs to be factored into the operation of every business. Training can be a formally organised process, as in a classroom or on-the-job practical training, but may be achieved by regular coaching as is evident in many apprenticeships.

Coaching is unlocking people's potential to maximise their own performance. It is helping them to learn, rather than specifically teaching them.

Coaching is a collaborative process, and when done correctly, it helps employees cultivate the confidence, skills, and thought processes they need to succeed in the workplace. A useful system for coaching is the GROW model:

- **G** set a Goal
- **R** examine the current Reality

- Explore the **O**ptions.
- **W** determine the Way forward

This can be a very effective process if used sensibly. The success depends very much upon who are being coached and their individual attitudes. Some will feel it patronising, while with others, it can be the ideal method of imparting knowledge.

15.48 BRIEFINGS AND DEBRIEFINGS

15.48.1 Briefings

A briefing is a discussion between two or more people using succinct information pertaining to a particular subject or task.

They should be used as a tool to improve teamwork, efficiency, and safety.

A good briefing should include the following points:

- Information pertinent to the task, i.e. location and timings
- Individual roles and responsibilities
- Safety-related information, e.g. safety equipment locations
- Relevant contact information and communication methods
- Desired outcomes
- Allowing questions from participants

There is a need, especially in UAS Operations, for a comprehensive briefing to take place prior to any task going ahead. This will ensure that all Remote Pilots, UA Observers, and all involved persons understand what is required of them for the duration of the operation.

Briefings are not just a tool for operational use, they are excellent when used to delegate a task or when initiating any projects.

15.49 RISK ASSOCIATED WITH A POOR BRIEFING

A poorly delivered briefing, or a briefing not delivered at all, risks a number of things. The operational objective or task may not get completed at all, not completed by a set deadline, or may be completed, but with incorrect end results.

Staff may end up not being where they are required, when they are required, or without the equipment that may be required.

The team may not understand what their roles and responsibilities are for the task or operation, what is expected of them, or have the correct lines of communication to use in the event of an emergency.

15.50 DEBRIEFINGS

If carried out correctly, a good debriefing is a crucial final step to any task or operation. It will assist with the prevention of repeating mistakes and is an excellent opportunity to look at improvement.

A debrief should take place after any significant task, operation, or project. If the task is a recurring event, a good debrief can provide valuable information to improve in the next instance, allowing for any lessons to be learnt.

The three basic categories to discuss are:

- What went well?
- What could have gone better?
- What could be improved for next time?

When discussing these categories, it will help to consider the following:

- What surprised you?
- Is there anything you know now that you wished you knew at the beginning of the task?
- Did anything unexpected happen? How was it resolved?
- How can those lessons be applied for next time?
- Resulting from this debrief, what can we change now?

15.51 PRACTICAL DESKTOP EXERCISE

An example of one such exercise I have run successfully on many occasions is offered next and is known as: "**OPERATION DESERT SURVIVAL**".

Some basic ground rules to assist with deriving maximum benefit include:

- Set a strict time limit for the exercise.
- Suggest to the group that they vote/appoint a chairman to oversee the proceedings and maintain discipline. He/She should make minimum contribution other than ensuring everyone can make their contribution.
- Nominate a secretary to take notes.
- Nominate a timekeeper to keep a check on events.

Expect to observe some of the following default behaviours from respective participants:

- Those who like the sound of their own voice and try to dominate the discussions.
- The shy retiring participant who likes to avoid participation.
- Those with strong practical suggestions.
- Those with totally impractical suggestions.
- Some behaving like naughty school children and try to highjack the process.
- Who are the natural leaders.
- Who lack the confidence to make suggestions.

15.51.1 How to Derive Additional Benefit from the Exercise

- Consider introducing an element of competition by dividing the group into two smaller groups.

- As an invigilator, throw a red herring into the process once or twice by changing some of the exercise criteria such as a reduced time available or a new contributory factor.
- Oversee a post exercise review and debrief by encouraging participants to analyse what they did, and how they think particular individuals' behaviour influenced the task.
- What went well?
- What went badly?
- If the exercise was to be repeated, what would they do differently?

15.51.1.1 Exercise Desert Survival, Team-Building Exercise

Instructor notes:

AIM: Develop communication, leadership, and negotiation skills.

Objective: Participants will have worked individually and as a group to establish the importance of listed objects and identified whether they perform better as an individual or a group. They will have recognised the aspects of teamwork involved and related this to the workplace.

The "Desert Survival" team-building exercise is a fun way to bring your group together as they problem solve, work as a team, and learn to trust one another.

Exercise:

Issue the **Scenario** and read through this with the participants.

1) Issue the **Score Sheet** and ask participants to first work individually to put items in order of importance and record in "My Ranking" column of the score sheet. They should not discuss their personal score at this stage.
2) They have ten minutes to complete this task.

Ideally if there are sufficient people in the class, split into two teams. Explain that they are the survivors of this accident.

Discuss what each group think are the attributes of a good team. For about ten minutes.

3) Without changing their individual "My Ranking" score, work with your team to arrive at a consensus of how the team rank the importance of which items are the most important.
4) Write the agreed order in the "Team Ranking" column of the score sheet.
5) They have 20 minutes to complete this task.

Remind the teams when there are five minutes left. When the time is up, bring the group back together again and discuss their answers.

Questions:

HAVE YOU DECIDED TO STAY OR GO?
What thought processes did they follow to address the problems?
What issues or questions did they consider?
In what order did they approach these issues?

How difficult was it to work out a team consensus?
Did you nominate a leader?
Did you write down or formalise a plan?
Did everyone finally agree?
Anyone did not agree with the team decision, how did they feel?

15.51.2 Scenario

15.51.2.1 "Desert Survival" Team-Building Exercise

It is 1:00 p.m. on a Saturday afternoon at the end of May. You and your teammates have just finished a two-day training in Casablanca, Morocco. You are all on board a chartered, twin-engine jet plane that is destined for Dakhla, Morocco, a small town on the coast of the North Atlantic Ocean, approximately 1,000 miles from Casablanca.

At the beginning of the flight, the captain came on the overhead speaker and invited you to sit back and relax during the two-hour flight. The first 50 minutes of the flight were fine.

Around this time, the Pilot comes back on the speaker to let you know that you are currently flying over the Sahara Desert and that weather reports showed a high temperature of 115 degrees.

Approximately one hour and ten minutes into the flight, you hear a loud blast, and the plane nosedives. Within minutes you realise that the cabin is losing pressure. When you look outside the windows, you notice that is desert below is growing larger as the plane rapidly descends toward the ground. You notice that the only things you can see out of your window are some large boulders and miles and miles of sand.

The Pilot informs you that the plane has blown an engine and is therefore, indisputably, going to make a forced landing, so all on board should prepare for a turbulent, possibly catastrophic landing somewhere in the desert. Within minutes, the planes crashes, and smoke and flames fill the cabin. All surviving passengers and crewmembers scramble to exit the plane.

Seven minutes after the crash, the plane explodes in a fiery ball that reduces it to rubble. With the exception of the airplane's captain and one crewmember, you, your teammates, one flight crewmember, and the co-captain have all survived the crash.

There are a large group of Oil Camps estimated to be about 120 nautical miles due east of the position where you think you are now stranded.

You must decide how to work together to survive the desert climate and terrain, get help, and hopefully make it out of the desert alive. On your way out of the plane, in the few minutes before it exploded, you and your teammates were able to salvage the items in the list below.

It is May and you and your teammates are dressed in business casual for the hot summer months of Africa. With only the clothes on your back and the items pulled from the wreckage, how will you survive?

The first question you must decide is do you stay together as a team at your current location or do you try to walk to the Oil Camps.

If you split the team up, who is going to take which of the saved items? (See the next section Exercise Desert Survival Answer Sheet.)

15.52 EXERCISE DESERT SURVIVAL ANSWER SHEET

ITEM	RANK	RATIONALE
Torch with four battery cells	4	Essential for night-time use
Folding knife	6	Cutting rope, food etc.
Air map of the area	12	To have idea of current location
Plastic raincoat (large size)	7	To collect dew overnight
Magnetic compass	11	Since awaiting rescue not much use
First-aid kit	10	Everyone is safe at present
45 calibre loaded pistol	8	For defence (3 shots recognised as distress signal)
Parachute (red and white)	5	Use as tent
Bottle of 1,000 salt tablets	15	Of no use in dessert
2 litres of water per person	3	A person needs about 4 litres per day in the dessert
A book entitled "Edible Desert Animals"	14	Food is less important than water in the dessert. Digestive process requires water.
Sunglasses for everyone	9	Protect against glare
2 litres of 180 proof liquor	13	Useful as antiseptic only. Causes dehydration if drunk.
Overcoats (for everyone)	2	Essential protection in dessert. Clothing helps ration sweat by slowing evaporation and prolonging cooling effect. Night-time temperatures can approach 0 degrees C at night.
A cosmetic mirror	1	Means of visually signalling.

FIGURE 15.1 Team building exercise.

Exercise Desert Survival Score Sheet

Item	My Ranking	Team Ranking	Answer	My Error	Team Error
Torch with four battery cells					
Folding knife					
Air map of the area					
Plastic raincoat (large size)					
Magnetic compass					
First-aid kit					
45 calibre pistol (loaded)					
Parachute (red and white)					
Bottle of 1,000 salt tablets					
Two litres of water per person					
A book entitled "Edible Desert Animals"					
Sunglasses for everyone					
2 litres of 180 proof liquor					
Overcoats (for everyone)					
A cosmetic mirror					
Score					

FIGURE 15.2 Master answer sheet.

Notes:

For an exceptionally fit team such as Special Forces, about 20–25 km per day is a realistic estimate of how far a group could travel in one day. On sand dunes, this could be reduced drastically to a few kilometres.

In summer, ground (sand) temperatures can reach about 30 degree C above the air temperature. (E.g.: Rub Al Khali's outside air temperature (OAT) in July is 58 degree C; therefore, sand temperature is 80 degree C). Try and sit about 12 inches above the ground minimum.

Night temperatures can plummet to freezing levels during the winter. It is possible to have a difference of about 50 degree C between about 16:00 and 04:00.

15.53 KNOWLEDGE CHECK

The following questions will give the reader the opportunity to check their progress and understanding of the previous chapter content prior to a review of learning outcomes.

1. List five competencies you think are essential to be an effective leader.
2. Analyse and list the different styles of management.
3. What do you understand by the GROW model?
4. Explain the mnemonic SMARTER.
5. What do you understand by the term active listening?
6. List five points to be included in a good briefing.
7. List three categories to be included in a good debriefing.
8. Describe what additional considerations should apply to a good debriefing.
9. Explain the issues that may arise from micromanaging.
10. Give three examples of how to motivate your team.

15.54 LEARNING OUTCOME REVIEW

LO 1: Understand the management needs for an Unmanned Aircraft Systems Operations Manager.

LO 2: Evaluate theories of management.

LO 3: Analyse how the practical application may affect own organisation's operations.

LO 4: Describe which management style you feel represents your own style.

Part Six

Practical Application in the RPAS (Drone) Industry

FIGURE PVI.1 Drone maintenance technician.

DOI: 10.1201/9781032620220-21

16 RPAS Operational Oversight

16.1 LEARNING OUTCOMES

By the end of the lesson, students will be able to:

LO 1: Investigate the structure and organisation competencies required as an Accountable Manager.
LO 2: Analyse the component parts of an approved Operations Manual.
LO 3: Describe the process to effectively update and maintain the Operations Manual.
LO 4: Understand the legal requirements for insurance coverage and data management.

16.2 OPERATIONAL OVERSIGHT

A part of the UAS Operation Manager's responsibility is general oversight of all factors associated with UAS operations. The UAS Operation Manager should be ensuring that all general operations are being carried out in compliance with the below subjects mentioned here:

- Operations Manual
- Operational Authorisation
- UK CAA Regulations

16.3 OPERATIONS MANUAL

A big part of their general oversight is the Operations Manual. The Operations Manual is created by the UAS operator and/or Accountable Manager, detailing policies and procedures for the safe operation of UASs in the specific category. The Operations Manual in its entirety should be complied with at all times. The UAS Operations Manager should ensure that all staff working on behalf of the organisation have received a copy of the Operations Manual and are happy to adhere to all information contained within. A non-exhaustive list of critical areas to ensure compliance is as follows:

- All crew that are operating on behalf of the organisation should have a copy
- It should be kept up to date at all times

DOI: 10.1201/9781032620220-22

- Not a yearly update of document.
- All crew to be informed and supplied with an updated document
- Planning and logs kept IAW the document
- All Operational RP should have their details contained, e.g. Flyer-IDs

16.4 OPERATIONS MANUAL UPDATE PROCESS

The Operations Manual must be updated whenever there is a change to the regulations or the organisation. Examples of these updates could be policies, procedures, management change, form amendments, and CAA CAP updates.

The typical update process will consist of:

1. Change notification, e.g. Skywise, Flight Safety Meeting minutes.
2. Evaluation Assessment of how the changes affect the organisation.
3. Implementation Update to the relevant documents.
4. Governance/approval, e.g. legal team or Accountable Manager's approval of change.
5. Publication amended document set is published within the organisation.
6. Distribution amended documents are disseminated to all appropriate personnel.

Here, we have an example timeline of what a typical Operations Manual update cycle will look like in Figure 16.1.

We will now break down the individual sections of the graph in Figure 16.2.

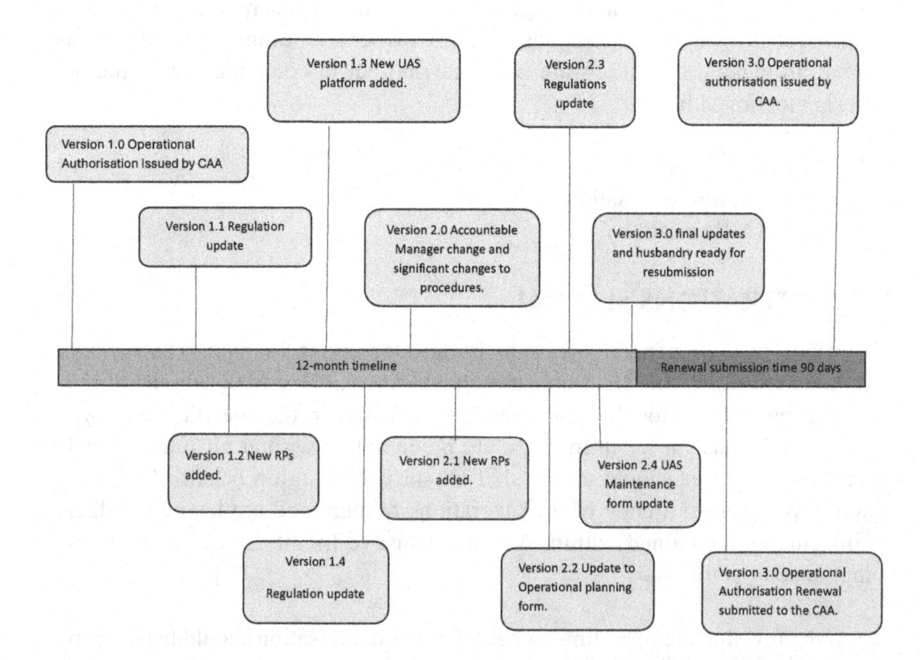

FIGURE 16.1 Block chart indicating a typical time cycle for Operations Manual updates.

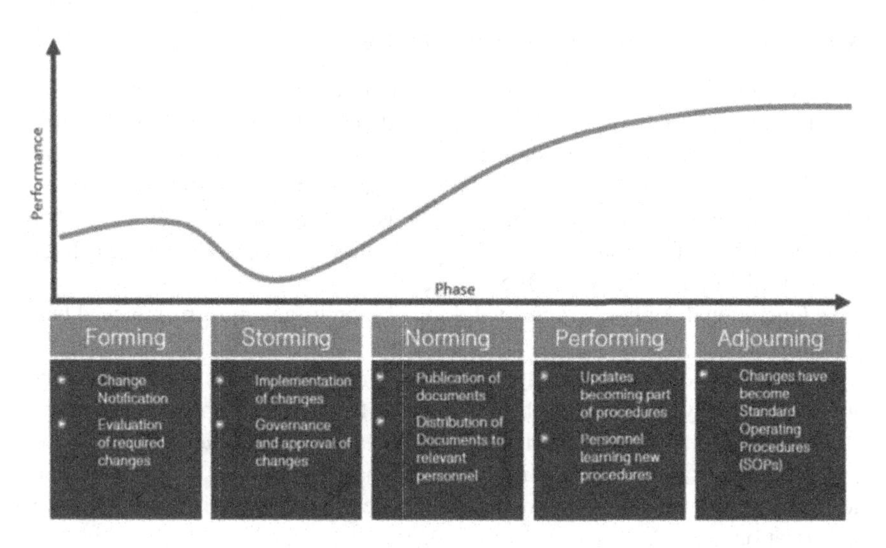

FIGURE 16.2 Phases of Team Development by Bruce W. Truckman.

Forming	This stage happens when the organisation receives a notification of change. This could be from any of the courses previously covered, such as the CAA service Skywise, internal notifications, or from the result of a Flight Safety Meeting. The UAS Operations Manager or equivalent will then evaluate the changes that are required.
Storming	This is where the documents are updated or changes are being implemented. Once complete, the document will go through an internal governance check or approval process—such as in the scenario when the UAS Operations Manager has made changes to the Operations Manual and the new version has gone to the Accountable Manager for approval before publishing. During this stage, you generally see a downward turn in performance among personnel as they are getting ready for a change.
Norming	This stage is said to occur where the updated documents have had approvals and are now published as the current working document. This is a critical stage, as the document also needs to be disseminated around the relevant personnel, ensuring that no one is referencing any old-document versions. This stage is where you start to see an upward turn in performance. This is due to personnel having published procedures to follow and are beginning to change any working practices.
Performing	This stage is where the updated documents are established, and the changes are starting to become easier to follow as the new working practices. It is critical at this stage to not introduce further changes as it will have a significant detrimental effect on performance.
Adjourning	This is the stage where any changes are now well practised and firmly part of any Standard Operating Procedures (SOPs).

16.5 OPERATIONAL AUTHORISATION

The Operational Authorisation is issued by the UK CAA, certifying that the UAS Operator can operate in the specific category, along with their operating limitations. The Operational Authorisation should be adhered to at all times, during all operations. All crew operating on behalf of the organisation should familiarise themselves with the copy contained within the Operations Manual. When received, it should be added to the Operations Manual. The Operations Manual Amendment Record should then be updated to reflect the change.

An Operational Authorisation is valid for 12 months and must be renewed before the expiry date.

Renewal can be applied for anytime during the three months prior to the expiry date. It is strongly advised that this is done at the earliest opportunity. Applying within the three-month period will not affect the anniversary date.

If the organisation has applied for an Operating Safety Case (OSC), any additional stipulates or amendments will be contained within the Operational Authorisation.

16.5.1 Insurance

UAS Insurance is a mandatory requirement for any commercial UAS Operation. The UAS Operator must ensure that suitable third party and equipment cover is in place. Next we have listed some of the other requirements of UAS Insurance.

16.5.2 Named Entity

The entity named on the Insurance Policy must be the same entity that is stipulated on the Operations Manual and Operational Authorisation.

16.5.3 EC Regulation 785/2004

Any UAS Insurance must comply with EC Regulation 785/2004. This is a requirement of all Air Carriers and Aircraft Operators.

16.5.4 Expiry Date

The Accountable Manager or UAS Operator must ensure that insurance is in place for all commercial operations and that it is renewed before the expiry date. If a lapse is found, all operations must cease until suitable insurance has been taken out.

16.5.5 CAA Regulations

The organisation must always adhere to the latest CAA Regulations. These changes will need to be updated in the Operations Manual whenever any changes are made. It is the responsibility of the UAS Operator to ensure that any changes are implemented immediately and updated throughout all relevant documentation. A good way to keep up to date is the CAA's Skywise Service.

http://skywise.caa.co.uk/

Once an account has been created with Skywise, the CAA will email through any relevant updates to changes (depending on what subjects you have asked for).

16.5.6 Data Management

All data gathered must be collected and stored in accordance with the UK Regulation (EU) 2016/679.

Breaches to this may result in significant fines. The Information Commissioners Office has some excellent information and resources regarding UASs.

All data management must be detailed within the Operations Manual and managed in accordance with the Data Protection Act (DPA 2018) and the General Data Protection Regulation (GDPR 2018).

While data management falls outside of the remit of the CAA, the UAS Operator must remain compliant with the other legislative or regulatory requirements for the acquisition, retention, and disposal of data.

The form on the next page is taken from CAP2378 Consultation Document for the Acceptable Means of Compliance and Guidance Material to Regulation (EU) 2019/947 as retained (and amended in UK domestic law) under the European Union (withdrawal) Act 2018. The following table can be found on pages 58 and 59 of CAP2378.

UASs have the ability to gather large volumes of sensitive date in a relatively short duration. Such data is stored in various locations, which are subject to data management policies.

- UA Flight Controller
- UA Internal Memory
- UA External Memory (Micro-SD)
- Command Unit (CU)
- Intelligent On-Screen Display (iOSD)
- Cloud Storage
- Cloud Data Processing Software

As a Data Controller, the UAS Operations Manager must ensure that all data gathered by a UAS is in compliance with their organisation's data protection policies, aligned to their data protection regulations.

This includes:

- Gathering
- Handling
- Retention
- Processing
- Transmission
- Disposal

1	**IDENTIFY THE PRIVACY RISKS THAT THE INTENDED OPERATION MAY HAVE**		
2	**DEFINE YOUR ROLE WITH RESPECT TO THE PERSONAL DATA COLLECTION AND PROCESSING.**		
	I am the (joint) data controller	I am the (joint) data processor	
3	**DATA PROTECTION IMPACT ASSESSMENT (DPIA)**		
	Have you assessed the need to perform a DPIA?	YES	NO
4	**DESCRIBE THE MEASURES YOU ARE TAKING TO ENSURE DATA SUBJECTS ARE AWARE THAT THEIR DATA MAY BE COLLECTED**		
5	**DESCRIBE THE MEASURES YOU ARE TAKING TO MINIMISE THE PERSONAL DATA YOU ARE COLLECTING OR TO AVOID COLLECTING PERSONAL DATA**		
6	**DESCRIBE THE PROCEDURE ESTABLISHED TO STORE THE PERSONAL DATA AND LIMIT ACCESS TO IT**		
7	**DESCRIBE THE MEASURES TAKEN TO ENSURE THAT DATA SUBJECTS CAN EXERCISE THEIR RIGHT TO ACCESS, CORRECTION, OBJECTION, AND ERASURE.**		
8	**ADDITIONAL INFORMATION**		

FIGURE 16.3 Template of Form describing how operational data is stored in accordance with GDPR regulations.

16.5.7 Freedom of Information (FOI) Act Requests

As a Data Controller, the UAS Operations Manager may receive an FOI request through their organisation-appointed FOI representative.

FOI is the right of any person to request any recorded information held by a public authority, which can include data from a UAS of such an organisation.

This can include but is not limited to:

- Flight Records
- Photographic or Videographic Data

- Operational Planning documentation
- RAMS

Appropriate advice can be obtained from the appointed FOI representative or from the Information Commissioners Office (ICO).

16.5.8 Oversight of Risk Management

The UAS Operations Manager and/or Accountable Manager must ensure that the Operational Risk Assessments are carried out in accordance with (IAW) the procedures contained in the Operations Manual. This will ensure that mitigations are not only appropriate but also being adhered to. This will also ensure that the RA process is not just a paperwork exercise.

The RP is immediately responsible for ensuring that the RA process is carried out sufficiently, but general oversight still remains with the UAS Operations Manager.

Any Risk Assessment done must be stored appropriately. This could be digitally or physically. All Operational Documentation must be stored for a minimum of three years, and the UK CAA can ask to see these documents during an audit.

Suitable management of risks will not only ensure compliance to the Operations Manual, Operational Authorisation, SMS, and HSE regulations but will also mitigate against financial and reputational risks.

16.5.9 UAS Operations Manager Compliance Checklists

A checklist is a useful tool to assist UAS Operational Managers ensuring that all operations are being carried out in accordance with legislation. The UAS Operations Manager should develop their own checklists so that they can be in line with their operational procedures. These checks can be further broken down into stages, to ensure compliance at each stage of the planning cycle. An example UAS Operations Manager checklist can be found in Figure 16.4.

16.6 OPERATIONAL RESTRICTIONS

The areas discussed in the above sections and table provide ideal placements that a UAS at his /her discretion.

Operations Manager can impose any operational restrictions. They could be in the form of:

- Specific or appropriate Risk Assessment Mitigations
- Additional RP flight hours for a complex operation
- Complex operating location familiarisation flights
- Keeping RPs to a limited number of UAS Operational timings
- Operational Aircraft Flight crew composition
- More stringent Operational Envelope for the selected UA

UAS Operations Manager Compliance Checklist			
Check to complete	**Details**	**Compliant Y/N**	**Signed**
Pre- Planning Stage			
Operation Location	Does RP have the correct Operation Location?		
Operational timings	Does the RP have the correct Operational timings? Day or night Ops? If night Ops, consider Human Factors.		
Travel	Does the RP have suitable transport for the operation? What is the travel time? Consider Human Factors.		
Accommodation	Is overnight accommodation required?		
Insurance check	Is insurance suitable and valid?		
RP availability	Are sufficient RPs available?		
RP currency	Are the selected RPs current and competent IAW Ops Manual?		
RP Fit to Fly	Are the selected RPs fit to fly?		
Aircraft selection	Is a suitable UAS available?		
Aircraft serviceability	Is the selected UAS serviceable? Are there enough UASs for the task?		
Operations Manual	Is the Ops Manual valid in date and does the RP have a copy?		
Operational Authorisation	Is the Op Auth valid? Does it cover the type of proposed operation?		
Ancillary equipment	Does the RP have the correct ancillary equipment for the task? E.g. Firefighting, First Aid		
PPE	Does the RP have suitable PPE for the operation?		
Risk Assessment	Has a suitable RA been carried out by RP? Are the mitigations appropriate?		
Communication Equipment	Does the RP have suitable communications i.e. mobile phone, radios.		
Contact Information	Does the RP have the relevant contact information for the operation? E.g. POC with the organisation, ATC, Emergency Services etc.		

FIGURE 16.4 Template of Form describing the components of a UAS Operations Manager's compliance checklist.

UAS Operations Manager Compliance Checklist			
Check to complete	**Details**	**Compliant Y/N**	**Signed**
On Site Stage (Spot Check)			
RP fit to fly	Is the RP currently fit to fly?		
Risk Assessment	Have the mitigations stated in the Risk Assessment been carried out?		
Operations Manual and Operational Authorisation	Does the RP have a copy or are they able to access a copy of the Ops Manual and the Op Auth?		
Safety Equipment	Is the RP utilising the correct Safety Equipment and wearing the appropriate PPE?		
Post Flight Stage			
Aircraft Serviceability	Is the Aircraft serviceable?		
Documentation Check	Ensure all operational documentation is completed and stored correctly IAW Ops Manual. E.g. Planning, RP Logbook, Maintenance Records, etc.		
Conduct Operational Debrief	What went well? What could have been done better? What could be improved for next time?		
Ops Manager Notes/Recommendations			
UAS Ops Manager Signature			
UAS Ops Manager Name			
Date			
Retention Date (Minimum of 3 years)			

FIGURE 16.5 Continuation from Figure 16.4 of a template of Form describing the components of a UAS Operations Manager's compliance checklist.

- Additional PPE requirements
- Specific Communication methods or equipment
- RP Fit-to-Fly requirements/Crew Health requirements

Additional restrictions can be imposed by the UAS Operations Manager at any stage of the planning cycle, even during the operation or during spot checks. However, adequate time and resources must be provided to any flight crew in order for them to impose these restrictions without having a detrimental effect on the operation itself.

16.7 LEARNING OUTCOME REVIEW

LO 1: Investigate the structure and organisation competencies required as Accountable Manager.
LO 2: Analyse the component parts of an approved Operations Manual.
LO 3: Describe the process to effectively update and maintain the Operations Manual.
LO 4: Understand the legal requirements for insurance coverage and Data Management.

16.8 KNOWLEDGE CHECK

1. List three required competencies required of an Accountable Manager.
2. List three of the main responsibilities of an Operations Manager.
3. List three Operational Restrictions which may be imposed by an Operations Manager.
4. Describe three criteria of an Operational Risk Assessment.
5. List five criteria that constitute part of the Operations Manual requirements.
6. List five criteria that form part of an Operations Manual update.
7. Who is responsible for Data Protection?
8. Where can you find the regulations for Data Protection?
9. Where can you find the regulations pertaining to what insurance coverage is operationally required?
10. Where can you obtain guidance relating to Freedom of Information responsibilities?

17 Understanding the Importance of UAS Maintenance

17.1 LEARNING OUTCOMES

By the end of the lesson, students will be able to:

LO 1: Understand the importance of Unmanned Aircraft System Maintenance.
LO 2: Analyse the importance of the maintenance process for Unmanned Aircraft Systems.
LO 3: Describe the steps of a maintenance process for an Unmanned Aircraft System.
LO 4: Evaluate Unmanned Aircraft System maintenance processes.
LO 5: Explain legal factors associated with the maintenance process.

A well-documented Aircraft Maintenance Record and schedule will greatly assist with managing an organisation's equipment, ensuring that is serviceable whenever it is needed. A set of comprehensive aircraft records will allow Remote Pilots to ensure that aircraft is serviceable at a glance. They should include as much detail as possible while still ensuring that the information being recorded is relevant. They are relatively easy to create and keep up to date but will allow an organisation to comprehensively track a range of useful information relevant to the safety of any flights. For example this could include trend information on battery cycles against flight times. Some UAS manufacturers have built in logs recording data such as battery usage, but it is recommended to have more detailed logs kept. This will also act as a backup of information in the event of any data loss resulting from issues during firmware updates. Currently, there is no airworthiness certification requirement for less than 25 kg UAS; therefore, the UAS operator is responsible for ensuring that the UA is airworthy prior to operation. To facilitate this the following may assist:

- Aircraft maintenance is the easiest when recorded in detail.
- There are good benefits for very little workload.
- It allows an organisation to track data and trend information on aircraft.
- Most newer drones will log most required information for you.
- It is recommended having personal backups in case of data loss on aircraft.

DOI: 10.1201/9781032620220-23

17.2 MAINTENANCE LOGS

Ideally, each aircraft should have its own folder/files with the following logs:

- Aircraft Flight hours
- Date of any servicing carried out
- Current firmware version and date of update
- Any component replacement
- Husbandry carried out, such as deep cleaning

All entries to the logs should have the date, name, and signature of the person who carried out the work and should be carried out by a suitably trained and competent person. It is also recommended that aircraft undergo a more thorough maintenance check periodically, especially if not used in for a period of time. These checks are in addition to standard start-up procedure checks. The following checklist details the basic requirements for the DJI Matrice 300RTK.

17.3 MOTORS AND PROPELLORS

- Rotation of the motors to check for any blockage or unnatural friction
- Propellor connection check to ensure a secure connection
- Propellor checks for any damage such as chips, cracks, or marbling

17.4 AIRFRAME, ARMS, AND ARM CONNECTIONS

- Looking for any damage, security of the arms, screws in place and secure
- Aircraft body is without any damage, all attached items are secure, and it is clean

17.5 LANDING GEAR

- Checking security of any legs, ensuring any attachments and screws are secured sufficiently.

17.6 BATTERY COMPARTMENT

- Checking any eject/button release mechanism is working correctly
- Ensuring batteries fit securely and that the connections are clean and without corrosion

17.7 VENTS

Ensuring any cooling vents are clean and clear from any obstructions.

17.8 RUBBER PORT COVERS

Checking security and cleanliness of any port covers.

17.9 GIMBAL

Ensuring gimbal is secured sufficiently and clean, and any dampeners are without damage or loose.

17.10 LENSES

- Checking any vision system/collision avoidance lenses are clean, secure, and without any cracks or damage
- Checking camera lenses are clean and without any damage

17.11 LIGHTING

Lighting is secure, without damage and working correctly carried out with the aircraft in a benign, low-risk environment (such as the A3 subcategory of the Open Category). This test flight should also be documented in the Maintenance Logs, including any relevant information, and be signed and dates by the RP who conducted the flight.

Next is an example of a recommended Maintenance Schedule published by DJI, for the Matrice 300 RTK.

17.12 AIRCRAFT MAINTENANCE RECORDS

Aircraft Maintenance Records can be as comprehensive as the organisation wishes, and more data is usually better. However, data recorded should be relevant. It will allow the organisation to track various information relevant to the use, such as total airframe flight time or when any parts have been replaced.

Next given are examples of the different recommendations for maintenance forms; however, this format does not need to be followed as it is for the organisation to create and maintain their own logs. There are some online services available that will create and store various forms such as RP Logbooks and Aircraft Maintenance forms; however, these vary in terms of quality and price, and you are then reliant on having internet connectivity and an active subscription to utilise.

Type	Maintenance Items	Maintenance Advice	Period
Basic Maintenance	1. Regular Maintenance Items 2. Updates and calibrations 3. Deep clean	Factory Service recommended	Total flight time is 200 hours or drone used for 6 months
Routine Maintenance	1. Regular Maintenance Items 2. Updates and calibrations 3. Deep clean 4. Component replacement due to wear and tear	Factory Service Mandatory	Total flight time is 400 hours or drone used for 12 months
Deep Maintenance	1. Regular Maintenance Items 2. Updates and calibration 3. Deep Clean 4. Component replacement due to wear and tear 5. Core component replacement.	Factory Service Mandatory	Total flight time is 600 hours or drone used for 18 months

FIGURE 17.1 The respective cycles of Basic, Routine, and Deep cycles of drone maintenance.

Maintenance Logbook													
Item	UA Category	UA Type	UA Manufacturer	UA Model	ID/Serial Number	Date Purchased	Maintenance Schedule	Operating Hours	Date of Maintenance / Repair	Work Conducted	Entity who conducted Work	Name of Engineer who is releasing the UA into Service	Notes

FIGURE 17.2 Form indicating the components of a drone maintenance cycle.

AIRCRAFT FIRST PARADE FORM

This form can be used as an initial checklist of the aircraft before heading out on an operation. It is usually carried out at HQ and can also be scheduled. For example every Monday or every first day of each month. Completing this will benefit the

AIRCRAFT FIRST PARADE RECORD									
AIRCRAFT DETAILS									
Manufacturer				Model					
Serial Number				Local Reference					
FIRMWARE									
Aircraft				RPS					
Batteries				Sensor					
AIRCRAFT									
Date of Check									
Firmware									
Frame/Shell									
Bolts									
Motors									
Landing Gear									
IMU									
Compass									
Vision System									
Collision System									
Infra-Red sensors									
Conspicuity Lighting									
Propellers									
CAA Operator ID									
Signature									

S	Serviceable	U/S	Unserviceable

FIGURE 17.3 Form used as an initial drone serviceability checklist.

organisation as it should identify any issues with the system before deploying on an operation.

COMMAND UNIT MAINTENANCE FORM

This should be used in a fashion similar to the "Aircraft First Parade Form". It is used to ensure that everything regarding the Command Unit is serviceable and fit for purpose.

SENSOR MAINTENANCE FORM

This form is a great way to track sensor serviceability and should be used in the same fashion as the "Aircraft First Parade Form".

COMMAND UNIT

Date of Check									
Firmware									
Body									
Bolts									
Sticks									
Switches									
Buttons									
Dials									
Connectors									
Battery									
Antenna									
Cables									
Signature									

S	Serviceable	U/S	Unserviceable

FIGURE 17.4 Form used for Command Unit Maintenance.

SENSOR

Date of Check									
Firmware									
Body									
Bolts									
Connection									
Lens									
Lens Function									
Ribbon Cable									
Gimbal									
Gimbal Clamp									
SD Card									
ND Filter									
Signature									

S	Serviceable	U/S	Unserviceable

FIGURE 17.5 Sensor Maintenance form.

17.13 BATTERY LOG

Batteries have a manufacturer-stated life span or maximum number of charge cycles. These forms are another way to record this information. They will also provide you with trend data, as you will likely notice a drop in flight times once the batteries approach and exceed the recommended cycle charge count. Many newer UASs use smart batteries. These smart batteries enable the flight software, such as the various DJI Flight Applications, to automatically log the charge cycle count. However, it is a good idea to keep a manual backup of this log as data could be lost with firmware updates.

Battery Logbook

Identifier	Manufacturer	Type	Smart/Dumb	Serial Number	Total Cycle Count	Date of last Maintenance	Maintenance Due	Any Outstanding Issues?

Maintenance Logbook													
Item	UA Category	UA Type	UA Manufacturer	UA Model	ID/Serial Number	Date Purchased	Maintenance Schedule	Operating Hours	Date of Maintenance / Repair	Work Conducted	Entity who conducted Work	Name of Engineer who is releasing the UA into Service	Notes

PDRA01 Operational Logbook – Required Headings

- Date
- UA Category
- UA Type
- UAS Serial Number
- Payload/s Type
- Fuel Type
- Fuel Grade
- Fuel Amount
- Battery Serial Number
- UA Centre of Gravity Calibrated?
- UA within MTOM?
- Total Flight Hours
- Next Maintenance Due
- Time until next UAS Maintenance
- Legal person releasing UAS into service
- Signature of legal person
- Date of release
- Remote Pilot Name
- RP Flyer-ID
- Start Location
- Start Time
- Landing Location
- Landing Time
- Total Flight Time
- Any occurrences or Airproxes?
- Any observations or issues?
- Any maintenance required?
- Any corrective actions required?

FIGURE 17.6 Battery Log indicating the number and times of charging history.

17.14 MULTIPLE AIRCRAFT

When an organisation has multiple aircraft of the same type, it is recommended that each system has its own identifier, as this will enable the UAS operator to track what equipment belongs to what UAS. This could be a serial number, name, or colour coding, and it should be used on all component parts such as command unit, sensors, external attachments, or batteries. Each individual aircraft should hold its own set of maintenance documentation, and this should be stored separately to avoid confusion. Forms can be kept either with the UAS as a hardcopy or electronically on the organisation's IT system.

17.15 LEGAL FACTORS ASSOCIATED WITH AIRCRAFT MAINTENANCE

UAS with a Maximum Take-off Mass (MTOM) greater than 25 kg and an Airworthiness Certificate is a requirement. To operate in the Specific Category, UASs must have a MTOM of less than 25 kg. This means that they do not require an Airworthiness Certificate. There are, however, other legal factors that the UAS operator must consider—mainly, Health and Safety Executive (HSE) requirements such as the Provision and Use of Work Equipment Regulations 1998 (PUWER).

In order to ensure that work equipment does not deteriorate to the extent it may put people at risk, employers, the relevant self-employed, and others in control of work equipment are required.[1] The level of maintenance and frequency of occurrence should be determined through the use of a Risk Assessment. This Risk Assessment should take into account:

- Manufacturer's recommendations
- Flight hours
- Operating environment
- Remote Pilot's experience of the system
- The risk from any foreseeable failure

Some safety critical components may require a more thorough and frequent level of attention. This can be programmed into a maintenance schedule.

Manufacturer's instructions should detail what maintenance is required to keep the UAS in a safe and serviceable state. Manufacturer's recommendations must always be followed.

17.16 PROVISION AND USE OF WORK EQUIPMENT REGULATIONS 1998 (PUWER)

PUWER places the responsibility for the upkeep of any work equipment on the people and companies who operate them or have control over them. This is regardless of if the equipment is owned by them or not.

PUWER requires that any equipment that is provided for work use is:

- Suitable for the intended use
- Safe to use, maintained, and inspected
- Used only by people who have had adequate information, instruction, and training
- Accompanied by suitable health and safety measures such as transport requirements for UAS and LiPo fire equipment for LiPo batteries
- Used only IAW the specific requirements, e.g. Flight Reference Cards, Operations Manual and Manufacturer's Recommendations

17.17 LEARNING OUTCOME REVIEW

LO 1: Understand the importance of Unmanned Aircraft System Maintenance.
LO 2: Analyse the importance of the maintenance process for Unmanned Aircraft Systems.
LO 3: Describe the steps of a maintenance process for an Unmanned Aircraft System
LO 4: Evaluate Unmanned Aircraft System maintenance processes.
LO 5: Explain legal factors associated with the maintenance process.

17.18 KNOWLEDGE CHECK

1 Who is responsible for oversight of all UAS aircraft Maintenance?
2. What checks are required for batteries?
3. What checks are required for sensors?
4. What checks are required for gimbals?
5. What checks are required for motors?
6. What checks are required for propellers?
7. What aircraft mass requires an Airworthiness Certificate?
8. If an aircraft does not require an Airworthiness Certificate, what other legal factors must be considered?
9. What are the limitations to operate in the Specific Category?
10. What does PUWER stand for?

NOTE

1 The above statement is extracted from **hse.go.uk.**

18 Organising a Flight Safety Meeting

18.1 LEARNING OUTCOMES

By the end of the lesson, students will be able to:

LO 1: Understand the requirements to organise and attend a Flight Safety Meeting.
LO 2: Analyse which personnel need to attend a Flight Safety Meeting and why.
LO 3: Discuss the function and structure of the Mandatory Occurrence Reporting Scheme (MORS).
LO 4: Understand the legal requirements for reporting and investigating.

Flight safety meetings are a critical part of any Safety Management System. They allow key members of an organisation to discuss subjects such as changes to the safety committee, Workplace Safety, Aviation Safety Policy, and the documentation of any miscellaneous matters. These areas can cover anything from items missing from a first-aid box, inadequate office lighting, to an incident during a flight operation. Flight safety meetings should be held at regular intervals, ideally once per quarter as an absolute minimum. Flight safety meetings are held to allow periodic review of all Flight Safety-related aspects. They are important to ensure that all key personnel of an organisation are kept up to date on safety-related aspects and provided a forum to raise any safety issues, investigate any incidents, apply corrective actions to any incidents, and allocate corrective actions to personnel. Flight Safety Meetings provide an excellent platform for disseminating critical information to the wider organisation through meeting minutes and any Stop Press or Pilots-to-See (PTS).

18.2 KEY PERSONNEL

The following personnel should be present in a Flight Safety Meeting.

- Accountable Manager
- UAS Operations Manager
- Flight Safety Manager
- Chief Pilot
- All available Remote Pilots
- Administrator

DOI: 10.1201/9781032620220-24

- Framework Provider Representative
- Communications Officer
- Compliance Monitoring/Quality Assurance Manager

The above list is not exhaustive. Any member of the organisation who may wish to raise a point can either go through their Line Manager and Air Safety Officer or be invited to attend the meeting. However, the above list is also not a minimum requirement. Your organisation may not have all these roles.

18.3 FLIGHT SAFETY MEETING AGENDA

The Flight Safety Manager or other person with a suitable role who is running the Flight Safety Meeting will be responsible for the meeting agenda. The agenda should be disseminated to the committee members prior to the meeting. This will allow them to prepare sufficiently. The agenda should state what subjects will be discussed and the points discussed under each area. Next, we have an example of a Flight Safety Meeting agenda.

Subject	Discussion	Remarks
Safety performance	➢ Review of events, accidents, and incidents	Safety Performance Indicators (SPIs)
Workplace Safety	➢ Fire safety ➢ First-Aid Equipment ➢ Environmental protection ➢ Manual handling ➢ Display screen equipment ➢ PPE ➢ Workplace-related injury	Head of Safety/Flight Safety Manager to present the Accident Book.
Aviation safety	➢ Review of reportable occurrences, accidents, serious incidents, or significant events ➢ Civil aviation publications/information notices ➢ Aircraft maintenance	
Training	➢ Staff safety training ➢ First aid at work representative ➢ Pilot competency	Safety Policy Safety Objectives ERP simulation Occurrence and hazard reporting
Policy and documentation	➢ Operations Manual/Operational Authorisation ➢ Amendment approvals ➢ Stop Press/Pilots-to-See ➢ Hazard log review	Amendment requests to be submitted to the Chief Pilot/ UAS Operations Manager no later than five days prior.
Safety performance	➢ Non-Conformance Reporting Review ➢ Compliance Monitoring Programme Schedule ➢ Human Factors	

FIGURE 18.1 Checklist used to highlight the agenda of a regular Flight Safety Meeting.

18.4 FLIGHT READINESS AND SAFETY REVIEWS

The UAS Operations Manager/Accountable Manager should periodically conduct Safety Reviews of completed documents such as Operational Planning, Maintenance Logs, and RP Logbooks. These post-action checks are another excellent opportunity to carry out the Operational Oversight checks discussed earlier in this publication.

18.5 TREND ANALYSIS

Safety reviews are an excellent opportunity to conduct in-house trend analysis. It would be worthwhile if the UAS Operations Manager reviews the areas explained in the next sections

18.6 SOCIAL MEDIA

You can often find information about safety-related issues posted on social media. This could be news articles, forums, or posts on sites such as LinkedIn.

18.7 OFFICIAL SOURCES

Official sources such as CAA Safety Notices, AAIB, and CHIRP publications will detail information about safety-related incidents; these could be reoccurring incidents that your organisation may need to be aware of, such as the DJI Matrice 200 series battery issue.

18.8 CAA SKYWISE

The official CAA service called Skywise is an excellent system for information. Once the user has signed up, email notifications will be sent out with any relevant information regarding subjects the user has selected.

Next is an example of a Skywise notification from 2019 regarding the DJI Matrice 200 series.

22nd March 2019 SW201&067 DJI Matrice 210 series drone
The CAA has received six Mandatory Occurrence Reports in the last 3 months affecting the DJI Matrice 210 series drone. The reports have indicated mat on each occasion me DJI M210 has malfunctioned resulting in rapid uncontrolled descent and consequent damage to me airframe upon impact with me ground. Reports suggest the issue may lie with one of the airframe's motors. We are working with the manufacturer and monitoring the situation.
All users of the DJI M210 series are advised to consider their obligations under Art, 241 of the ANO and avoid flying over people or property until further notice.
SW2019/067
Category: Drones

FIGURE 18.2 Skywise notification from 2019 regarding the DJI Matrice 200 series.

18.9 MANDATORY OCCURRENCE REPORTING SCHEME (MORS)

Mandatory occurrence reporting has been a part of the fabric of UK aviation operations since 1976 to support the continued learning from aviation occurrences. Occurrence reporting in the UK is mandated by UK Regulation 376/2014 which requires the reporting of safety-related occurrences involving UK airspace users. The purpose of occurrence reporting is to improve aviation safety by ensuring that relevant safety information relating to civil aviation is reported, collected, stored, protected, exchanged, disseminated, and analysed. It is not to attribute blame or liability. All airspace users are required to report MORS to the UK CAA, with the exception being annex one aircraft, where reporting is voluntary (although strongly encouraged by the UK CAA). Occurrence Reports are treated confidentially to maintain full and free reporting from the aviation community and to protect the identity of the individual. What should you report? In a nutshell, any occurrence that you feel could have an impact on aviation safety should be reported, as this will ensure that we always review and learn from events. Mandatory Occurrence Reporting in the UK is outlined in UK Regulation 376/2014 with additional guidance presented in implementing regulation 1018/2015.

18.10 COMPLETING A MORS

Reporters should submit MORS to the UK CAA in the ways explained in the next sections.

18.11 AVIATION REPORTING PORTAL

The aviation reporting portal can be found at: https://aviationreporting.eu/. From here, reporters can submit Occurrence Reports to the UK CAA (and EASA member states) using the online reporting forms or offline pdf forms. The reporting portal is designed to allow reports to be submitted from individuals, and further details about how reports can be submitted using the reporting portal can be found in CAP1496.

18.12 KEEPING FLIGHT SAFETY MEETING MINUTES

After the Flight Safety Meeting has taken place, it is critical to follow up on any points raised. The meeting minutes and any other relevant documentation, such as a Pilots-to-See or a Stop Press, should be distributed to all relevant personnel in detail, who are allocated to head up any required actions that have resulted from the meeting. Any documentation that has been amended or created during the meeting, such as the agenda, minutes, or PTS/Stop Press, should also be filed in accordance with the organisation's policy. Meeting minutes are essentially a written record of what has been discussed during the Flight Safety Meeting. As there is likely to be a considerable amount of information discussed during a Flight Safety Meeting, comprehensive meeting minutes are essential to ensure that all relevant points have been logged. Meeting minutes should record and remind of subsequent activities and assigned responsibilities, detail the accountability of any tasks, and ensure the dissemination of critical information.

In Figures 18.3 and 18.4, we have an example of a completed Flight Safety Meeting Minutes form.

Flight Safety Meeting Minutes

In Attendance (Name and Role)		Apologies (Name and Role)	
UAS Operations Manager Chief Remote Pilot Flight Safety Manager Framework Provider Representative		Accountable Manager Communications Liaison Officer	
Minute	Subject	Discussion Points	Action, Responsible Person, and Conformance Date
1	Changes to Safety Committee	• Apologies from the Accountable Manager as they are unable to attend due to sickness. • Apologies from the Communication Liaison Officer unable to attend due to Annual Leave. • The Flight Safety Manager would like to congratulate Joe Bloggs on the promotion to Chief Remote Pilot.	N/A N/A N/A
2	Workplace Safety Examples: ➢ COVID-19 measures ➢ First-Aid Equipment ➢ First-Aid Training ➢ Office Equipment ➢ Transport ➢ General Points	In line with Government Policy change, the Accountable Manager has now stipulated that the wearing of masks is no longer compulsory in the workplace, but it is still recommended. The UAS Operations Manager has informed the committee that the First-Aid equipment annual check is complete. The Chief Remote Pilot has informed the committee that all Remote Pilots require refresher training of their First Aid at Work certificates as they expire by 01/06/2022. The UAS Operations Manager has informed the committee that the company fire extinguishers are due their annual serviceability checks. The UAS Operations Manager has stated that the MOTS of the company vehicles are to be renewed. Due to expire on 01/08/2022. The Flight Safety Manager has received multiple complaints within the workplace due to the lighting in the offices. They have advised that the bulbs are to be replaced.	N/A Chief Pilot to check by 12/08/2022. Flight Safety Manager to organise by 31/05/2022. Chief Pilot to check by 12/08/2022. Chief Pilot to organise by 12/08/2022. UAS Operations Manager to organise by 31/05/202.

FIGURE 18.3 Suggested content to be included in Flight Safety Meeting Minutes.

In Attendance (Name and Role)		Apologies (Name and Role)	
3	Aviation Safety ➢ Examples: ➢ Reportable Occurrences ➢ Accidents ➢ Serious Incidents ➢ Aircraft Maintenance ➢ Batteries ➢ Training	Occurrence Report- Airprox when training with 2 × company drones. No injuries or equipment damage. RPs require additional training on airspace de-confliction. Occurrence Report—RP came into contact with moving propeller suffering lacerations. Required safety training on airmanship principles; in addition, RPs require refresher course on First Aid at Work. The Chief Pilot advised that our DJI Phantom 4 is approaching 500 flight hours, and it would be advisable to take it offline to have a service done. The UAS Operations Manager has advised to send it to heliguy for a full service. The UAS Operations Manager has advised that he will require a replacement aircraft while DJI P 4 is having a service. Approval will be asked from Accountable Manager to purchase a DJI Mini 3 and add to service fleet. The Chief Pilot has advised the committee that 3 TB50 batteries are swollen. They are currently quarantined and awaiting safe disposal. The UAS Manager has recommended we replace the 3 TB50 batteries at the same time that we order a new Mini 3 and will seek approval of the Accountable Manager. The Chief Pilot has advised that when we receive the new Mini 3, all RPs will require familiarisation flights with the UA before any commercial flights are carried out.	Chief Pilot to organise by 31/05/2022. Chief Pilot to organise airmanship training by 30/05/2022. Flight Safety Manager to organise First Aid Training by 12/06/2022. Chief Pilot to organise by 31/05/2022. UAS Manager to organise—liaise with Chief Pilot to ensure replacement here before DJI P 4 goes in for service. Chief Pilot to organise by 20/05/2022. UAS Manager to order DJI Mini 3. Chief Pilot to arrange upon delivery of Mini 3.

FIGURE 18.3 (Continued)

In Attendance (Name and Role)		Apologies (Name and Role)	
4	Policy and Documentation ➢ Operations Manual ➢ Operational Authorisation ➢ Pilots-to-See/Stop Press	The UAS Manager has reminded the committee members that the Operations Manual was updated to Version 2.7 dated 05/06/2022. This was to add new RPs to the approved register of RPs. The new document has been approved and disseminated. Please ensure that all old versions are replaced. Operational Authorisation is due for renewal on 18/09/2022. The UAS Operations Manager has advised that all documentation should be checked for compliance as the application will be submitted asap into the 3-month renewal window.	All personnel
5	Any other business	The UAS Operations Manager has advised the committee that heliguy released a Blog about "UK Drone Laws 2022" early in May. They have asked it to be disseminated to all operational personnel as it will be good for a quick refresher of the regulations. It also contains information on Operating Safety Cases. This is pertinent as we may be moving down this route. www.heliguy.com/blogs/posts/uk-dronelaws-where-can-I-fly	All Operational Personnel

FIGURE 18.3 (Continued)

The date of the next Flight Safety Meeting is scheduled for 12/08/2022 at 10 am. Any questions relating to the content of this document should be directed to the undersigned in the first instance.

Signature	xxx
Full Name	John Hancock
Role within the organisation	Flight Safety Manager
Date	12/05/2022

DISTRIBUTION LIST

Accountable Manager		UAS Operations Manager	
Chief Remote Pilot		Flight Safety Manager	
Framework Provider Representative		Communications Liaison Officer	
Remote Pilot 1		Remote Pilot 2	

FIGURE 18.4 Continuation of the template of suggested content to be included in Flight Safety Meeting Minutes.

RECORD RETENTION
Not for destruction before 5 years (60 months) from date of publication.

18.13 PILOTS-TO-SEE/STOP PRESS

A great idea for ensuring documents are disseminated around key personnel is to create a "Pilots-to-See" folder. This is also referred to as a Stop Press. These documents should detail any updates or changes to areas such as:

- Internal document updates
- Policy documents
- External legislative changes
- Risk Assessment mitigations
- Any safety-related incidents

One way to ensure compliance is to have all relevant personnel sign a compliance sheet to state that they have read and understood any entries. It should also contain contact information for the relevant organisation's point of contact if they have any queries.

18.14 MEETING MINUTE DISTRIBUTION

The following personnel should be included on the distribution list to receive meeting minutes.

- Accountable Manager
- UAS Operations Manager
- Flight Safety Manager
- Chief Pilot
- All Remote Pilots
- Administrator
- Framework Provider Representative
- Communications Officer
- Compliance Monitoring/Quality Assurance Manager

18.15 LEARNING OUTCOME REVIEW

By the end of the lesson, students will be able to:

LO 1: Understand the requirements to organise and attend a Flight Safety Meeting.
LO 2: Analyse which personnel need to attend a Flight Safety Meeting and why.
LO 3: Discuss the function and structure of the Mandatory Occurrence Reporting Scheme (MORS).
LO 4: Understand the legal requirements for reporting and investigating.

18.16 KNOWLEDGE CHECK

1. Who should attend a Flight Safety Meeting?
2. How often should you convene a Flight Safety Meeting?
3. Why does the meeting require the recording of minutes?
4. What does MORS stand for?
5. What is Skywise?
6. Who should be included in the distribution of Flight Safety Meeting minutes?
7. What information should be included in Pilots-to-See/Stop Press?
8. What is CHIRP?
9. What does the AAIB stand for?
10. What is a Flight Safety /Readiness review?

19 Flight Readiness Review

19.1 INFORMATION SOURCES

The UAS Operations Managers may be responsible for operational oversight of proposed UAS operations. These can originate from a variety of sources, including, but not limited to:

- Internal approved Remote Pilots
- Internal non-approved Remote Pilots
- Internal departments
- External Entities/Framework Providers
- Media and PR General Public Factors

UAS Operations Managers may be responsible for issuing authorisation to take place from land or infrastructure under the control of the organisation; therefore, appropriate governance should take place. While external organisations will operate under their own qualifications or Operational Authorisation, due diligence remains the responsibility of the hosting organisation. Factors to consider include:

- Operating category qualifications
- Competency and currency
- Operational planning and RAMS
- Insurance vetting
- Work permit site induction and PPE
- Escort status
- Compliance and competency

Where the UAS Operations Manager is not competent in evaluating UAS specific Operational Planning documentation/RAMS, an appropriate representative should be appointed for assurance. A comprehensive evaluation of the aforementioned factors should take place prior to operational approval being granted. It should be recognised that each UAS operator will have differing policies and procedures than that of the hosting organisation; therefore, due diligence in the form of a Flight Readiness Review provides assurance that they are commensurate with and comply with the host organisation's SMS.

DOI: 10.1201/9781032620220-25

19.2 NEGATIVE INDICATORS

Some common factors to be aware of when conducting a Flight Readiness Review includes:

- Incomplete documentation
- Inaccurate or irrelevant information
- Inconsistent or contradictory information
- Delays in the distribution of documentation
- Legacy or outdated information
- Failure to comply with RP currency or qualification expiry
- Expiring or expired Operational Authorisation
- Falsified documentation

20 Future of Unmanned Aircraft Systems

This very short section highlights some of the future developments in the RPAS Industry reported by Mark Blaney who is HELIGUY (Accountable Manager) and a front-line operator responsible for maintaining company standards. His role requires him to be aware of all current Industry standards and gives him the ideal platform to maintain an awareness of future developments.

20.1 EVLOS/BVLOS

Beyond Visual Line of Sight (BVLOS) operations are increasing in popularity. The current method for flying BVLOS is to apply to the UK CAA for an exemption to an Operational Authorisation by using an Operating Safety Case (OSC). Currently, BVLOS is only able to be flown through the use of an OSC application and must be flown in atypical/segregated airspace. However, if the current trend of popularity continues, BVLOS operations will be the future of the majority of UAS Operations.

20.2 UAS TRAFFIC MANAGEMENT

Currently, all segregated UAS airspace is through Danger Areas. Generally, UASs will have freedom of operations within the bounds of the atypical/segregated airspace although some constraints may be imposed when applying for use. Until UASs can comply with requirements for flights in non-segregated airspace, occasional or one-off BVLOS flights may be allowed through Temporary Danger Areas (TDAs). When applying for these flight permissions, it takes into consideration procedures for areas such as emergency procedures, loss of control data link, and collision avoidance with other aircraft. Some flights may be permitted outside of established Danger Areas, but applications for these operations will be taken on a case-by-case basis by the UK CAA and will likely involve the use of Temporary Danger Areas (TDAs). As UAS technology is evolving rapidly, and businesses are pushing the uses of UASs for a range of purposes, the requirement for more established airspace is growing rapidly. This is largely down to organisations wanting to use UASs for package or medical deliveries and Emergency Services having the requirement for longer range operations. Future UASs will likely require a greater level of technology on both Ground Control Stations and Aircraft than what is currently being used for standard commercial operations. UASs will likely require some form of Secondary Surveillance Radar

DOI: 10.1201/9781032620220-26

Transponder (SSRT). These devices constantly transmit a code or beacon that will alert local Air Traffic Control or other nearby aircraft to the UAS position, bearing altitude, individual identifier, and any emergency status.

Any BVLOS operations will require the Remote Pilot to be able to immediately take control of the UASs and respond to any Air Traffic Control requests in the same timeframe as what could be expected from any manned aircraft.

Other future UAS requirements are likely to be an approved method of assuring terrain clearance to ensure that the UAS is well clear of any higher terrain etc., and a method for them to comply with IFR/VFR (Instrument Flight Rules and Visual Flight Rules) appropriate for the class of airspace being flown in.

20.3 FLIGHT MANAGEMENT SYSTEMS

There is a growing need for robust Flight Management Systems (FMSs). Various companies have released or are currently developing FMSs. These are an excellent tool for a UAS Operations Manager to organise their Remote Pilots, fleet management, and operational oversight. They facilitate storage and sharing of mission planning between Remote Pilots and the Accountable Manager/UAS Ops Manager.

20.4 MULTI-ROTOR VERSUS FIXED-WING

As you will be aware, UAS Operations require a good degree of planning. Aircraft selection is a very important factor in planning for a mission, as multi-rotor and fixed-wing have very different capabilities. Due to the differences in their strengths and weaknesses, consideration should be taken when picking the right aircraft for the mission.

For example an operation that would require a large area to be flown and mapped would require multiple flights from a multi-rotor with a typical flight time of approximately 15 to 35 minutes, compared to a fixed-wing aircraft that can typically fly for approximately 45 to 180 minutes per flight. The reason for this is that fixed-wing aircraft fly a lot more economical than a multi-rotor. As a case study, a 5 km × 7 km area would take between 20 and 30 flights plus from an average multi-rotor to map, whereas an average fixed-wing aircraft could do this in approximately 5 flights, only being restricted by VLOS flight distances of 500 m from the RP. However, if an operation would require smooth video, or still images of an object not directly below the aircraft, then a multi-rotor will obviously be the most relevant choice due to its ability to hover in position. If a mission would require a 3D model of a structure, a multi-rotor again would be the best choice due to being able to get images from a greater angle than directly below the aircraft.

20.5 MULTI-ROTOR VERSUS FIXED-WING COMPARISON TABLES

Multi Rotor Strengths	**Multi-Rotor Weaknesses**
Ability to hover	Short flight times
Extremely manoeuvrable	IP ratings
Stabilised video	Not aerodynamic
Indoor operations	
Proximity flying	
Relative quick mission planning	
Small TOL area requirement	
Fixed-Wing Strengths	**Fixed-Wing Weaknesses**
Long flight times	Large TOL required, VTOL (Vertical Take-off and Landing) Fixed-Wing systems are increasing in popularity and availability
Large area coverage	Constantly in motion
Mostly autonomous flights	Greater degree of planning required
Mostly IP rated	
Very aerodynamic—can glide if loss of power	

FIGURE 20.1 Strengths and weaknesses of fixed wing compared with multi-rotor devices.

21 Preparing an Operational Safety Case

Your organisation may be required require to apply for a **Drone Operating Safety Case** to operate outside Standard Permissions.

An Operating Safety Case (OSC) is a complex, three-volume operations manual that is submitted to the Civil Aviation Authority (CAA) and enables Pilots to operate outside of the confines of a Standard Permission—formerly a PfCO, now an Operational Authorisation. An OSC enables drone operators to push the boundaries of their operations to expand the potential and effectiveness of their missions and provide a competitive edge.

21.1 WHAT IS AN OSC?

A three-volume Operations Manual for complex operations, an OSC (Operational Safety Case) is submitted to the CAA and enables Pilots to operate outside of the parameters of UKPDRA01 within the Specific Category.

An OSC is required for numerous operational procedures, such as:

- Flying less than 50 m from uninvolved people
- Flying less than 50 m from uninvolved buildings/property
- Flying less than 50 m horizontally from crowds of people
- Flying more than 400 ft in altitude
- Flying beyond visual line of sight (BVLOS)

The aim of the OSC is to present sufficient evidence that all relevant hazards and resultant safety risks have been identified within an operation and that these safety risks have been reduced to a tolerable and As Low As Reasonably Practicable (ALARP) level. This ensures that the required operational safety requirements have been met and best practice is adopted by drone operators in the UK.

21.2 HOW TO GET AN OSC FROM UK CAA

Prerequisites and required volumes are as follows:

Operators are required to submit a bulk of information to the CAA to obtain an OSC.

DOI: 10.1201/9781032620220-27

For the majority of applications, this information needs to be presented across three volumes. These volumes are:

- **Operations Manual:** Comprehensive summary covering aspects such as the type of operation, Emergency Response Plan, operational procedures, and company safety policy.
- **UAS Systems:** In-depth submission about the aircraft(s) being used.
- **Safety Risk Assessment:** Submission of detailed information about the safety risks associated with the operation.

Your organisation may find that an efficient method to obtain an OSC is to contract the task to a specialist consultant.

heliguy who have contributed very closely to the compilation of this publication are able to offer such a service. Contact them via the link as given in Part Eight.

22 Understanding the Auditing Process for UAS Operators with Typical Auditing Questions

22.1 OVERVIEW

Regular Auditing of an Organisation is an integral part of maintaining a Safety Assurance and Compliance programme. Audits can be:

- Internal Company Audit
- External Audit Companies
- Regulatory Audits (e.g. CAA)

The philosophy of auditing has changed over the years. Initially, many organisations were fearful of "being found out" for non-compliance and behaved in an illogical way by trying to hide many of the areas where modifications or corrective actions should take place. Auditors were viewed as being enemies who were there to close down an organisation if they could.

Nothing could be further from the truth. Today, most auditors are only too keen to help organisations attain the appropriate legal standards for operations.

Many even publish and send an advanced set of questions to companies relevant to the area of operation they intend to scrutinise. Sensible operators try to keep regular oversight and update of their procedures and processes to avoid a panicked last-minute review and reorganisation because, "The auditors are coming"! The following Specimen Auditing Questions represent examples of the type of questions an organisation needs to be able to address to satisfy some areas of Safety and Compliance.

22.2 TYPICAL AUDITING QUESTIONS

Organisational Elements and Requirements

Is the manual system presented in a format which can be used without difficulty?
Is there a correct table of contents?
Are the record of revisions/amendments, as well as the list of effective pages available, and/or the list of effective chapters updated?
Are all pages numbered throughout the manual?

DOI: 10.1201/9781032620220-28

Does every page or chapter provide information about the effective date and the revision status?
Is there an annotation of page layout?

Electronic Data Processing

Are there concepts and procedures for documents in an EDP solution?
Is the accessibility as well as the usability defined and are respective procedures available?
Is the backup system defined and the reliability ensured?
Are there common provisions in regard to physical security?

Structure of the Management System Documentation

Is an overview available over the manuals, which are in place to comprehensively define the lines of responsibility and accountability as well as the organisation's key processes?
Is there an introductory text describing the scope and applicability?
Are all relevant chapters of the SMS systematically structured?

Including:

Log of revision
List of effective chapters or pages
Highlights of latest amendment
Table of content
Abbreviations, terms, and definitions

System of Amendment and Revision

Is there a comprehensive amendment procedure, valid either for the whole Management System?
Documentation or for the individual manuals/parts, as specifically required?
Is there a reference to the applicable system of amendment and revision?
Is there a log of revisions/amendments and a record of revision highlights?
Is there a list of effective pages/list of effective chapters?
Are the different types of revisions which may be carried out defined (standard revision, temporary revision, urgent revision)?
Is there a statement that revisions/amendments are to be processed and concluded entirely before new changes are initiated?
Does this revision/amendment procedure ensure compliance verification prior to the submission of the document to the CAA?

Changing Elements Requiring Prior Approval

Does/do the procedure(s) consider both kind of changes: Changes needing prior approval and changes not needing prior approval by the competent authority?
In the case of revisions/amendments not impacting elements requiring prior approval, is there a statement that the compliance manager ensures, that no element requiring prior approval is included?

For changes requiring prior approval—does/do the procedure(s) consider to conduct a safety Risk Assessment to be provided to the competent authority upon its request?

Does the operator have processes for the production of manuals and any other documentation required and associated amendments?

Is the operator capable of distributing operational instructions and other information without delay?

Change Elements Requiring Prior Approval

Is there a process which covers elements requiring prior approval or is it included within the system of amendment and revision?

Is there a reference to the "Compliance List" or "List of Acceptance and Approvals"?

Does the amendment procedure consider that the application needs to be submitted before any changes take place?

Does this revision/amendment procedure ensure compliance verification prior to the submission of the document to the CAA?

Does this amendment procedure ensure that the application for the amendment of an operator/organisation certificate should be submitted at least 30 days before the date of the intended change?

Does this amendment procedure ensure that in case of a planned change of a nominated person, the operator should inform the competent authority at least 20 days before the date of the proposed change?

Does this amendment procedure ensure that unforeseen changes should be notified at the earliest opportunity, in order to enable the competent authority to determine continued compliance with the applicable requirements and to amend, if necessary, the operator certificate and related terms of approval?

Is there a statement that amendments requiring prior approval may only be implemented upon the receipt of a formal approval?

Change Elements NOT Requiring Prior Approval

Is there a comprehensive procedure which defines the handling of elements not requiring prior approval?

Does/do the procedure(s) differ between changes requiring prior approval and changes not requiring prior approval by the CAA?

Does the procedure include guidance on how to distinguish between changes requiring prior approval and changes not requiring prior approval?

Is there a statement that revisions/amendments are to be processed and concluded entirely before new changes are initiated?

Does this revision/amendment procedure ensure compliance verification prior to the submission of the document to the CAA?

Is there a requirement that the application needs to be submitted before any amendment/revision takes place, even for those elements not requiring a prior approval?

Is there a statement that the compliance manager ensures that no element requiring prior approval is included?

Does the procedure include that the revision/amendment is to be submitted to the CAA at least 30 days prior to the planned publication?

Document Control of External/Foreign Documents

Is there a process for the amendment of defined foreign documents, which lists the different kind of external documents, the responsible persons, and their activities?
Are updated versions of the relevant documents available?
Organisational and Strategic Planning
Is there a statement regarding the organisation's Vision, Mission, and Values?
Is the statement consistent with the Safety Policy?

Scope of Activity

Are the different types of activities clearly defined (either in the SMS or in other relevant manuals)?
Are the privileges and detailed scopes of activities defined for which the organisation seeks certification or is certified for?
Are the privileges and detailed scopes of activities defined according to what the operator declared and/or is seeking authorisation or is authorised?
Are the scopes of activities defined and consistent with the terms of any approval(s) held?
Does the operator provide a description of the proposed area and type of operation, including the type(s), number of aircrafts to be operated, and the respective specific approvals?
Does the organisation provide a list of approved training courses and—if already approved—the number of the certificates?

Statement of Complexity

Is the organisation defined, either as non-complex or complex?
Is this categorisation appropriate?
Relevant Legal Requirements and Standards
Is a list of relevant legal requirements/regulations existing?
Did the organisation provide the competent authority with a statement that all the submitted documentation was verified and was found in compliance with the applicable requirements?

Compliance Statement

Is there a statement, signed by the Accountable Manager, which confirms that the organisation will continuously work in accordance with the applicable requirements and the organisation's documentation as required by the respective Annexes?

Alternative Means of Compliance (AltMoC)

Are there provisions related to AltMoCs?
Does the assessment of Alternative Means of Compliance include a demonstration that the implementing rules are met?
Does the assessment of AltMoCs include a Risk Assessment?

Is the list in conformity with any granted approvals?
Is there a statement that the organisation must not implement AltMoCs without having received the formal approval?
Is there a list and brief description of approved AltMoC(s)?

Location, Facilities, and Infrastructure

Is there list containing a general description of the location and the facilities?
Is there a clear reference to another manual including the respective chapter if the definition is made outside the SMS?
Are appropriate ground-handling facilities available to ensure the safe handling of flights?
Are there operational support facilities at the main operating base, appropriate for the area and type of operation?
Is the available working space at each operating base sufficient for personnel whose actions may affect the safety of flight operations?
Are appropriate flight operations accommodation (facilities) available?
Are appropriate facilities for theoretical knowledge instruction available?

Access and Power of Authorities

Is the power and access for authorities specified?

Organisation, Lines of Responsibilities, and Accountabilities

Is there a general description of the organisation including an organigram of all related companies?
Are the titles and functions of nominated persons and management personnel defined?
Does an organisation chart exist which shows the lines of responsibility between the Accountable Manager, the Safety Manager, the Compliance Monitoring Manager, and nominated persons?
Is there a cross-reference table, providing the link to subordinated organisational structures in other manuals, as required by the available organisation's approvals?
Is a Safety Review Board (SRB) designated?
Is a Safety Action Group (SAG) designated?

Management Personnel Names and Contacts

Is there a list of all management personnel?
Are there references to the lists of nominated personnel as required by the specific subparts?
Are contact details of management personnel provided? Duties, Responsibilities and Accountabilities
Is the accountability in each description of the duties and responsibilities clearly specified?
Are the responsibilities and duties comprehensively defined (including their contribution to an effective Safety Management and Compliance Monitoring)?

Accountable Manager

Are the accountability, responsibilities, and duties of the Accountable Manager comprehensively defined?
Is a direct safety accountability of the Accountable Manager including the responsibility for establishing and maintaining an effective Safety Management System defined?
Do the defined responsibilities include the endorsement of the safety policy?

Safety Manager

Are the responsibilities and duties of the Safety Manager comprehensively defined?
Are the responsibilities within the organisation defined for hazard identification, Risk Assessment, and mitigation?

Compliance Monitoring Manager

Are the responsibilities and duties of the Compliance Monitoring Manager comprehensively defined?
Has the Compliance Monitoring Manager direct access to the Accountable Manager?
Is it ensured that the Compliance Monitoring Manager is not one of the nominated persons?

Safety Management

Is there a comprehensive safety policy including all relevant elements defined?
Is the Safety Policy endorsed by the Accountable Manager?
Does the Safety Policy reflect organisational commitments regarding safety and its proactive and systematic management?
Is there a statement indicating that the sole purpose of safety reporting and internal investigations is to improve safety and not to apportion blame to individuals?
Does the Safety Policy reflect and foster the Just Culture?
Is the Safety Policy promoted and deployed with visible endorsement throughout the organisation?

Hazard Identification and Risk Management

Is there a process to identify aviation safety hazards entailed by the activities of the organisation?
Is there a process to evaluate and manage the associated risks of the hazards?
Is there a process for the implementation of corrective and preventive actions in order to mitigate the risk and verify their effectiveness?
Is the hazard identification and safety risk management integrated into the day-to-day activities of the organisation?
Is it ensured that contracted activities are subject to hazard identification and risk management?
Are the responsibilities for hazard identification and Risk Assessment defined?
Is there a process for reactive and proactive hazard identification?

Is there a risk management process addressing the analysis of hazards in terms of likelihood and severity?
Are the levels of management who have the authority to make decisions regarding the tolerability of safety risks specified?
Is there a process to control and mitigate the associated risks of a hazard to an acceptable level?
Are there examples existing for reactive and proactive hazard identification, Risk Analysis, and mitigation action?
Are internal safety investigations performed beyond the assessment of the occurrences?
Is there a process by which the safety performance is monitored and verified in comparison to the safety policy and the organisation objectives?
Is the Risk Management process periodically reviewed and improved?
Is an Emergency Response Plan established?
Is there a documented process to manage safety risks related to a change?
Does this process enable to identify external and internal changes that may have an adverse effect on safety?

Safety Review Board (SRB)

Is a Safety Review Board designated?
Is the Safety Review Board chaired by the Accountable Manager?
Are the SRB responsibilities and duties comprehensively defined?

Safety Action Group (SAG)

Is a Safety Action Group designated?
Are the SAG responsibilities and duties comprehensively defined?

Change Management

Is there a documented process to manage safety risks related to a change?
Does this process enable to identify external and internal changes that may have an adverse effect on safety?

22.2.1 Safety Review Board (SRB)

Is a Safety Review Board designated?
Is the Safety Review Board chaired by the Accountable Manager?
Are the SRB responsibilities and duties comprehensively defined?

Safety Performance Monitoring

Are safety objectives defined?
Are the safety objectives communicated to all employees?
Are Safety Performance Indicators (SPIs) defined?
Is the Safety Performance checked on a regular basis?
Is the Safety Performance communicated to all staff on a regular basis?

Safety Promotion and Communication

Is there a procedure on how employees are informed about safety issues?

Safety—Studies—Reviews–Surveys– Investigation

Is a process defined how internal safety studies are conducted or how external studies are considered?
Is a process defined how internal safety investigations are conducted?
Is a process defined which addresses proactive and reactive evaluation of facilities, equipment, documentation, and procedures?

Reporting Scheme

Is a Reporting System defined?
Is a feedback process integrated within the Reporting System?
Is there a statement that the overall purpose of the reporting scheme is to improve safety performance and not attribute blame?

22.2.2 Compliance Management

Is the organisational set-up of the Compliance Monitoring adequate to the size, complexity, and activity of the organisation?
Is the scope of the Compliance Monitoring appropriate to the complexity and activity of the organisation?
Are the elements of the Compliance Monitoring Programme complete?

Occurrence Reporting

Is there a link to the occurrence reporting procedures as required by the specific organisation?
Is there a possibility to report anonymously and confidentially?
Is there a reference to the reporting portal www.aviationreporting.eu "Aviation Safety Reporting" of the EU for mandatory occurrence reporting and, desirably, for voluntary reporting?
Do the procedures ensure that any occurrence, serious incident, and accident are reported by the organisation to the CAA and, in case of serious incident or accident, to the AIIB?
Do the procedures include defined time frames for each reporting step and stipulate that reports shall be made available to the competent authority as soon as possible but latest within 72 hours if the individual is directly reporting to the authority?
If the individual is reporting to an organisation, the report shall be submitted within 72 hours to the organisation and within another 72 hours from the organisation to the authority.
Are specific forms provided as required by the organisation and its terms of approval?
Are the reports processed as defined in the Reporting and Feedback System?
Do the procedures ensure that the corrective and the preventive actions to avoid similar occurrences in the future are reported to the competent authority?
Are all Occurrence Reports retained and stored regardless of their significance?

Audit and Inspections

Is there a procedure for planning and scheduling the programme of audits?
Does the audit procedure include the verification of practical samples?
Is there a procedure for planning and scheduling the programme of inspections?
Are scopes/areas specifically defined for the audit and inspections?
Is there an audit plan or audit plans covering the relevant elements and audit-scopes?
Do the audits also focus on the integrity of the organisation's Management System including safety management?
Does the audit and inspection procedure include the reporting, as well as the follow-up and corrective actions?
Are audit- and inspection reports available?
Do they provide all relevant information?
Is a sample provided?
Does the audit and inspection procedure include the supervision of the implementation of corrective actions and the monitoring of their effectiveness?
Are the audit- and inspection processes applied, practised, and effective?

Auditors and Inspectors

Is it ensured that inspections and audits are carried out by personnel not responsible for the function, procedure, or products being audited?
Do all the auditors and inspectors have relevant knowledge, background, and experience as appropriate to the activities being audited or inspected, including knowledge and experience in Compliance Monitoring?
Is there a list of auditors by name and inspectors by function?

Findings, Corrective, and Preventative Actions

Are there provisions and procedures related to the handling of findings?
Does the procedure require that a corrective action plan is defined which addresses the effects of non-compliance, as well as its root cause?
Does the procedure require that any safety measures mandated by the CAA are implemented?
Does the procedure require that any relevant mandatory safety information issued by the agency including airworthiness directives are implemented?
Does the procedure require that the implementation of measures and its effectiveness is monitored?

Classification of Findings

Is a categorisation established to classify findings according to their severity?
Are there time limits allocated for the completion of corrective actions/measures?

Leasing

Is there a process describing how to handle lease agreements?
Is there a description that ensures that the operator of the lease-in aircraft is not subject to an operating ban?

Is the documentation of the application exhaustive in relation to the respective lease arrangement?
Is there a process describing how equal level of safety is reached for third-country operators?

Contracting and Leasing

Does the introductory text include a statement, indicating that the organisation ultimately remains responsible for the products or services provided by the contractor?
Is a process defined which ensures that the contracted or purchased services and/or products do conform to the applicable requirements and, where applicable, that the contractor holds the required certificates and approvals?
Does the process include the verification of the contractor's compliance with the defined philosophies, policies, procedures, and requirements of the organisation?
Does the process include the verification of the contractor's facilities and resources and show the competence of the contractor to execute the contracted tasks?
Is there a list containing the contracted products or services including the contact details of the contractor?
Does the CAA have access to the contracted organisation in order to determine compliance with the applicable requirements?
Does the organisation have a written agreement with the contractor?
Are the contracted activities clearly defined?
Is it assured that the contracted activities are subject to Compliance Monitoring and safety management?
Depending on the product/service provided, is it assured that contractors are trained on the defined philosophies, policies, procedures, and requirements of the organisation?
Are contractors, if applicable, provided with the organisation's documentation or parts thereof?
Does the process provide details on actions to be taken, should a contractor product or service fall below requirements—initiation and monitoring of corrective/preventive actions?

Record Keeping and Archiving

Does the organisation have a system of record that allows storage and reliable traceability?
Are all records accessible and available?
Is it specified how the records are kept (hardcopies or software)?
Are the records safeguarded?
Are the retention periods defined?
Is a list available with all necessary documents and their retention periods?

Management Evaluation

Are the Safety Performance Indicators integrated in the Management Evaluation?
Is the Management Evaluation performed on a regular basis?

Continuous Improvement

Is a process defined which addresses proactive and reactive evaluation of facilities, equipment, documentation, and procedures?

Is a process defined on how results out of evaluations are used to improve the system?

Emergency Response Planning

Is there a statement regarding the scope and objectives of the Emergency Response Planning (ERP) concept?

Is there a documented process ensuring an orderly and safe transition from normal to emergency operations and return to normal operations?

Does the ERP concept outline a communication and notification plan, including communication and notification to the authorities and the emergency response team?

Are the composition, role, and contact details of the emergency response team defined?

Are guidelines and initial response procedures for the emergency response team members defined so that the initial tasks may be performed correctly?

Are the actions to be taken by the organisation or specified individuals in an emergency defined?

Is the initial set-up of required facilities such as the Crisis Management Centre defined?

Is there a procedure regarding restrictions of crew scheduling after a serious incident or accident?

Is there a procedure regarding safeguarding and retaining of relevant data and records such as FDR and CVR recordings, training and checking results, technical records, and flight planning documents?

Is there a documented process on how to notify the CAA, including relevant numbers and contact details?

Is the ERP concept integrated in the Organisation's Management System?

Is the ERP concept ensuring safe continuation of the operations during the emergency?

Is the ERP concept coordinated with the Emergency Response Plan of other organisations such as the home base airport or code share partners?

Does the ERP concept address a public health emergency or pandemic as well?

Management System Training

Is a Management System Training concept defined?

Does the concept consider the requirements of all of employee levels and functions?

Does the concept consider initial and continuous training?

Basic Training of All Employees

Does the basic training include all fundamentals of the organisation's Management System?

Does the basic training ensure that all employees are aware of their responsibilities?

Is the Management System training adequately integrated within the training and checking programme for flight crew and, if applicable, for technical crew?

Is the Management System training adequately integrated within the staff training programme?

Management Training (Advanced)

Does the advanced training consider the requirements of all of the management personnel levels and functions?
Does the advanced training ensure that all management personnel are aware of their responsibilities?

Continuous Management System Training

Is there a Continuous Management System Training defined?
Is the Continuous Management System Training based on a systematic analysis of factual data and results derived from the Safety Management, Compliance Management, Reporting- and Feedback System, and Management Evaluation?
Is the Continuous Management System Training adequately integrated within the training and checking programme for flight crew and, if applicable, for technical crew?
Is the Continuous Management System Training adequately integrated within the staff training programme?

Part Seven

Research

FIGURE PVII.1 DJI Mavic 3 which has landed on a tree stump in a woodland.

DOI: 10.1201/9781032620220-29

23 DRONOTS

23.1 DRONOTS (DRONE NON-TECHNICAL BEHAVIOURS)

heliguy and the Applied Psychology and Human Factors Group based at Aberdeen University are currently investigating the development of a Behavioural Marker System known as DRONOTS for use in the RPAS Industry.

This system is following principles similar to the NOTECHS system used extensively in aviation and a new evolution of that system known as HELINOTS for use specifically in the North Sea Offshore Industry and the Search and Rescue sectors.

A brief summary of two key potential research options and associated objectives are outlined next.

23.1.1 Objectives and Associated Methodology

23.1.1.1 Objective 1: Identification of Core NTS and Associated Influencing Factors

The research team proposes a study to identify the core NTS and associated elements utilised by drone operatives. This will be conducted within a systems approach which will also identify the potential factors influencing NTS performance in this context. All members of the research team have a significant expertise in this human factors area, having conducted work across similarly high-risk sectors.

As such, specifically, the team proposes to:

- Interview a selection of drone operatives (approximately 12 interviews, with 6 lone operatives and 6 team based, and more may be required depending on data saturation and range of mission types) who operate across a range of mission types.
- Use a task analysis technique, and probing semi-structured questions, to better understand the different Pilot contexts, typical actions, and key NTS.
- Utilise content analysis as a method of qualitative data analysis to identify tacit knowledge regarding NTS utilisation; describe core NTS categories, elements, and behaviours; and assess influencing factors.
- Develop a preliminary NTS framework, outlining core NTS and associated elements, relative to each mission type, including assessment of similarities and differences across roles.
- Develop a preliminary framework of potential performance-influencing factors.

DOI: 10.1201/9781032620220-30

23.1.1.2 Objective 2: Refinement of NTS Drone-Specific Framework

The customary second stage for the development of an NTS framework, and associated behavioural marker system, consists of additional data gathering and refinement. Typically, this will encompass data triangulation via a survey, followed by discussion groups or Delphi method to confirm elements and behaviours.

- Construct a survey designed to assess attitudes towards NTS within the drone industry, including the perception of the most important and frequently utilised skills (derived from the preliminary framework described before). The mixed measures survey will enable the analysis of the key behaviours associated with NTS, the confirmation of the most important aspects, and further investigation of the potential influencing factors. Minimum sample size for such a survey would be 100 participants, equally split across different drone flight types.
- The survey data would be evaluated using quantitative and qualitative analysis techniques to confirm the key NTS categories, elements, and behaviours as well as the most important influencing factors.
- The survey could be combined with discussion groups (or discussion groups could be conducted instead of the survey) whereby small groups (four to five participants) of drone industry experts and Pilots would be gathered to discuss the preliminary NTS framework described above, in order to ensure that the framework terminology is appropriate, refine the elements and behaviours, and provide examples of good and poor behaviours where further details are needed.
- The methods above would enable the construction of the first DRONENOTS framework.

Additional avenues for research:

Assessment of DRONENOTS framework/marker system: Commonly, the application of Behavioural Markers is assessed via the use of simulation or video. Specific scenarios are constructed and presented to a range of experts/Pilots whose role is to identify and score (good/poor) the perceived NTS. A test of inter-rater reliability is then conducted to determine the reliability and validity of the tool and identify any areas of further development/tailoring.

Objective evaluation of situation awareness: Given the vital nature of situation awareness in aviation, it is important to gather objective (not self-report) data regarding the utilisation of this skill. Potential methods for further exploration of this skill include eye-tracking, sabermetrics, and general physiological methods. Similar to the above, these methods are commonly utilised as part of a simulated mission in order to assess changes in eye movements/effort/workload.

Project outputs: In each of the above cases, the research team will deliver a detailed academic report outlining the findings of the analysis. This report will outline the steps taken to develop the relevant method, a summary of the data gathered and a series of recommendations for next steps.

These reports can be used as a basis to structure CRM-style training modules in human factors to the sector, as it will provide the necessary understanding of the NTS utilised on the sharp end. As detailed previously, it is now overwhelmingly accepted across academia and industry that safety-critical behaviours are part of a wider context, and this research will characterise the application of NTS within that specific context.

Proposed by:
Dr Oliver Hamlet and Dr Amy Irwin
Applied Psychology and Human Factors Group

24 Using the COM-B Method for Influencing Behaviour Change in Training

24.1 THE COM-B MODEL OF BEHAVIOURAL CHANGE

Capability—defined as the individual's psychological and physical capacity to engage in the activity concerned. It includes having the necessary knowledge and skills.

Motivation—defined as all those brain processes that energise and direct behaviour, not just goals and conscious decision-making. It includes habitual processes, emotional responding, as well as analytical decision-making.

Opportunity—defined as all the factors that lie outside the individual that make the behaviour possible or prompt it.

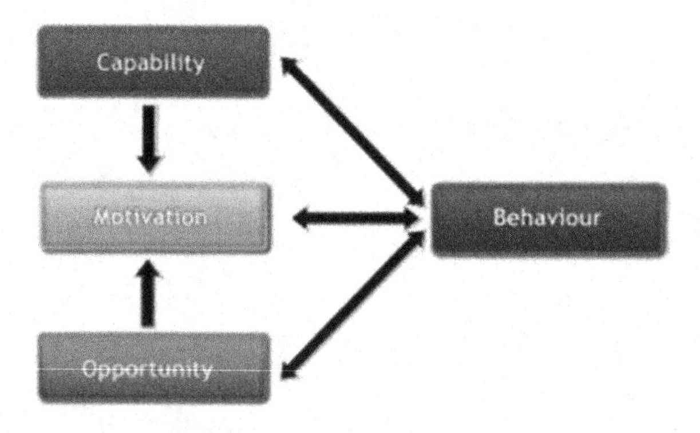

FIGURE 24.1 COM-B model of behavioural change.

DOI: 10.1201/9781032620220-31

24.2 THE BEHAVIOUR CHANGE WHEEL

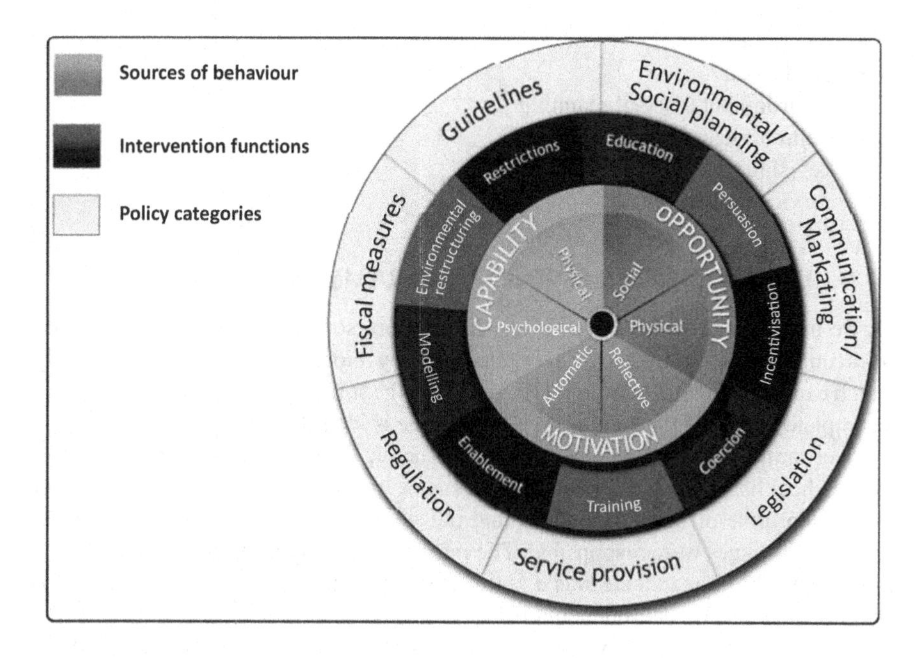

FIGURE 24.2 COM-B behavioural change wheel.

24.3 SOURCES OF BEHAVIOUR

This is exactly the correct way to present the COM-B model as defined by the inventors of the system.

- **Capability**—Physical and psychological
- **Motivation**—Automatic and reflective
- **Opportunity**—Social and physical

24.4 INTERVENTION FUNCTIONS

- Education
- Persuasion
- Incentivisation
- Coercion
- Training
- Enablement
- Modelling
- Environmental restructuring
- Restrictions

24.5 POLICY CATEGORIES

- Regulation
- Fiscal measures
- Guidelines
- Environmental/social planning
- Communication/marketing
- Legislation
- Service provision

24.6 DEVELOPING TRAINING PROGRAMMES USING COM-B

As we investigated in the **Part Four—Education,** sometimes the learning journey from unknown to known can be interrupted by blockages or misunderstandings which can be difficult to identify. Occasionally, the training programme will come to a complete halt until the reason for the blockage is identified.

Over many years' experiences as an instructor, I have often wished there was a tool or checklist to help me isolate the blockages. In my opinion, most of the problems were often quite difficult to identify and the solutions not always obvious. Usually, behavioural issues were responsible. The introduction of systems such as NOTECHS Behavioural Marker scheme was a fantastic developmental milestone to actually be able to classify various influencing behaviour traits.

However, how to actually change some of those behaviours once identified was another problem. While working with the APHF Group in Aberdeen, I was introduced to the COM-B method used in public health and thought that it could be adapted very easily and used in the context of helping identify and mitigate some of the teaching and learning problems encountered by several disciplines of employees in the Aviation Industry.

As with NOTECHS, using COM-B can provide an excellent starting point for further analysis if not always able to directly provide a definitive solution.

I continue to develop this theory and hope to be able to recommend it to assist with certain drone-training issues.

24.7 EXAMPLES OF USING COM-B FOR TROUBLESHOOTING STUDENT ISSUES

24.7.1 Scenario (Quite a Common Occurrence in Aviation)

Student A has been transferred to a brand-new aircraft type which is much more complex than his currently operated type with higher demands upon IT skills to understand and manage the control systems.

Even after extra instructions, he is still struggling to operate the systems with an acceptable level of confidence. He is obviously very upset and frustrated at not being able to achieve the required company standard. The company recommendation is to achieve the relevant standard after about 50 hours of training. After an extra five hours of training, he is becoming more depressed and has reached a plateau of competence.

24.7.2 Company and Instructor Considerations

The company has already invested £80,000 in his training on the previous type and a further £25,000 on the new type training. His training records reflect that his performance on the previous type was acceptable but often minimum standard.

24.7.3 Options

- Continue with the training programme but with a different instructor (possible success)
- Remove him from the course (waste of finances but necessary)
- Return him to previous type (always a difficult option due to the behaviour resulting from stigma of failure and impact upon future career development)
- If depressed behaviour persists, terminate his employment (often comes with legal and contractual issues and sends a message to other employees about employee care).

This is a complicated decision and requires a sensitive analysis of what action is necessary and how the company will approach the issue.

24.7.4 Conclusions

Very often, such situations are badly handled because those in charge do not have the ability or the time to make an objective analysis.

24.7.5 Using COM-B to Assist with Analysis

24.7.5.1 Capability

- Did the student ever have the capability to be transferred to a new type?
- If no, what processes failed to identify his potential issues?
- Did anyone study his training records adequately before the transfer?
- If he was assessed as not capable, what aspects of his ability would suggest he would have problems?
- Are there any aspects of the current training programme, which impact upon a student's capability?
- Are there any other students facing similar issues?
- What are the actual logjams in his performance impeding his progress?
- At what stage of the training did his performance fail to meet the standard?
- Was the student lagging behind from the beginning?
- Are there any social issues impacting his capability?
- Are there any personal issues impacting his capability?
- Are there any financial issues impacting his capability?

24.7.5.2 Opportunity

It would appear that the student has taken advantage of the opportunity to join the course, but maybe it raises some questions for the training department.

- How robust was the selection process for the new type?
- Who was involved in the selection process?
- What criteria were considered?

24.7.5.3 Motivation

As a general rule, aircrew are rarely short of motivation, and their personalities are very similar to that of professional sportsmen in terms of self-confidence and competitiveness. They have a natural inclination towards continued professional development. When these traits are not present, it is usually an indication of some other underlying issue.

Questions for the training department are as follows:

- Is the student motivated to succeed on the course?
- Did he want to be on this course?
- Have we investigated to find out that if he has any issues which temporarily affect his motivation?
- Are there any social issues influencing his motivation?
- Are there any personal issues influencing his motivation?
- Are there any financial issues influencing his motivation?
- Are there any company employment issues such as potential redundancies influencing his motivation?

Part Eight

Resources, Links, and Recommendations

DOI: 10.1201/9781032620220-32

The following pages contain many useful sources of information for those who have a deeper interest in researching the content of this publication.

They have been organised into:

1. Chapter-specific reference documents
2. General publications of interest
3. Online resources
4. Video library
5. Regulatory publications

CHAPTER-SPECIFIC REFERENCE DOCUMENTS

Part One

Chapter 1—Safety Management Systems

International Civil Aviation Organisation (ICAO)—ICAO Doc 9859

www.icao.int/SAM/Documents/2017-SSP-GUY/Doc%209859%20SMM%20Third%20edition%20en.p

EASA—ORO.GEN 200

www.easa.europa.eu/en/the-agency/faqs/orogen

US Federal Aviation Administration (FAA) entitled Safety Management System Fundamentals

www.youtube.com/watch?v=RDJZ2S-bkm8)

Chapter 2—Safety Policy and Objectives

UK CAA—Just Culture

www.youtube.com/watch?v=ugl9FaBOlKA

Flight Safety Organisation

A Roadmap to a Just Culture: Enhancing the Safety Environment

https://flightsafety.org/files/just_culture.pdf

Health and Safety Executive (HSE)

www.hse.gov.uk/

Chapter 3—Safety Risk Management

Mandatory Occurrence Reporting (MOR)

www.caa.co.uk/our-work/make-a-report-or-complaint/report-something/mor/occurrence-reporting/

Voluntary Occurrence Reporting (VOR)

www.caa.co.uk/our-work/make-a-report-or-complaint/report-something/mor/occurrence-reporting/

European Co-ordination Centre for Accident and Incident Reporting Systems (ECCAIRS)

https://aviationreporting.eu/

Confidential Human Factors Incident Reporting Programme (CHIRP)

https://chirp.co.uk/aviation/

HSE—Provision and Use of Work Equipment Regulations 1998 (PUWER)
www.hse.gov.uk/work-equipment-machinery/puwer.htm
European Helicopter Safety Team (EHST)
www.easa.europa.eu/community/topics/ehest-heritage-useful-sms-tools
HE:5

Part Two

Chapter 6—Human Factors, Crew Resource Management, Team Resource Management

Airbus Industries
A Statistical Analysis of Commercial Aviation Accidents 2023
www.airbus.com/sites/g/files/jlcbta136/files/2023-03/A-Statistical-Analysis-of-Commercial-Aviation-Accidents-2023.pdf
***Safety at the Sharp End: A Guide to Non-Technical Skills* by Rhona Flin, Paul O'Connor**
CRC Press, 15 May 2017—Technology and Engineering
***Crew Resource Management* by Barbara G. Kanki (Editor), José Anca (Editor), Thomas R Chidester**
"Just a Routine Operation"—Captain Martin Bromiley
www.youtube.com/watch?v=JzlvgtPIof4
Railway and Transport Safety Act 2003—Chapter 20, part 5—Aviation; Alcohol and Drugs
www.legislation.gov.uk/ukpga/2003/20/contents

Information Acquisition and Processing

Federal Aviation Administration—Airport Safety Information Video Series
www.faa.gov/airports/safety-video-series
www.youtube.com/user/FAAnews/video
Selective Attention Test
www.youtube.com/watch?v=vJG698U2Mvo
fMRI scan
www.youtube.com/watch?v=rJjHjnzmvDI
National Library of Medicine—Reflexology under fMRI
www.ncbi.nlm.nih.gov/pmc/articles/PMC3668141/
The Brain
https://youtu.be/kMKc8nfPATI
Hippocampus
www.youtube.com/watch?v=GDlDirzOSI8
Left and Right Hemispheres
www.youtube.com/watch?v=TQ51Gsb98ec
Neuroplasticity
www.youtube.com/watch?v=J8wW1t1JqUc
Cerebral Cortex
www.youtube.com/watch?v=mGxomKWfJXs

Cerebral Cortex Overview
www.youtube.com/watch?v=X-m0JDCw6TE
Structure of the Nervous System
www.youtube.com/watch?v=jmD0LBdAvlE
Functions of Nervous System

Decision-Making

European Helicopter Safety Team (EHST)
www.easa.europa.eu/community/topics/ehest-heritage-useful-sms-tools
www.youtube.com/watch?v=qrK-FBdjGk4

Professor James Reason—Generic Error Modelling System (GEMS)
https://skybrary.aero/articles/generic-error-modelling-system-gems

Workload Management

Risk Factors in the Technique for Human Error Rate Prediction (THERP) (Swain and Guttmann 1993).
https://digital.library.unt.edu/ark:/67531/metadc828716/

Stress and Stress Management

American Institute of Stress (AIS)
www.stress.org/stress-effects
www.stress.org/wp-content/uploads/2011/11/50-Common-Signs-and-Symptoms-of-Stress.pdf
COPE Scotland
www.cope-scotland.org/

Personality, Cultural and Generational Differences

Personality Test
www.16personalities.com

European Helicopter Safety Team (EHST)
www.easa.europa.eu/community/topics/ehest-heritage-useful-sms-tools
HE: 8

Monitoring and Intervention

UK CAA Videos Monitoring Matter
Low Energy Management
www.youtube.com/watch?v=qWMaXwnU5g0&feature=player_embedded
Vertical Flight Path Management
www.youtube.com/watch?v=gmUIpR9hWD4&feature=player_embedded
Fault Management
www.youtube.com/watch?v=JQ7aTt4ws7U&feature=player_embedded
Near Stall
www.youtube.com/watch?v=TLW5Sr0B-7w&feature=player_embedded
Level Bust 1
www.youtube.com/watch?v=IKI0oR2DUs0&feature=player_embedded

Level Bust 2
www.youtube.com/watch?feature=player_embedded&v=4nvXZwmxRb4
Level Bust 3
www.youtube.com/watch?v=o5uVuTvknSU&feature=player_embedded
GPWS
www.youtube.com/watch?feature=player_embedded&v=zMY42bb8jUs

Automation Philosophy and Use

European Helicopter Safety Team (EHST)
www.easa.europa.eu/community/topics/ehest-heritage-useful-sms-tools
HE 9: Automation and Flight Path Management

Surprise and Startle Effect

Fight or Flight
www.youtube.com/watch?v=m2GywoS77qc

Assessment and Behavioural Markers

Part Three

Chapter 13—Accident/Incident Investigation/Case Studies

Air Accident and Investigation Branch (AAIB) www.gov.uk/search/all?keywords=+Drone+accidents+2023&order=relevance
https://aaib.dronedesk.io/
AAIB Drone Accident Reports
www.gov.uk/aaib-reports?keywords=drones
2023 AAIB Annual Review
https://assets.publishing.service.gov.uk/media/6604534991a320001182b13d/AAIB_Annual_Safety_Review_2023.pdf
Maritime Safety Agency
www.gov.uk/government/organisations/maritime-and-coastguard-agency
RSSB (Rail Safety and Standards Board)
www.rssb.co.uk/safety-and-health/improving-safety-health-and-wellbeing/understanding-human-factors
GOV.UK
www.gov.uk/government/publications/investigating-accidents-to-unmanned-aircraft-systems/investigating-accidents-to-unmanned-aircraft-systems
HSE Accident Investigation
www.hse.gov.uk/pubns/books/hsg245.htm
Cranfield University
www.cranfield.ac.uk/Courses/Short/Transport-Systems/Fundamentals-of-Accident-Investigation
UK CAA
Bowtie Risk Analysis
www.caa.co.uk/Safety-initiatives-and-resources/Working-with-industry/Bowtie/About-Bowtie/Introduction-to-bowtie

Part Four

Chapter 14—Teaching and Learning

For an in-depth understanding of the principles of modern-day teaching and learning practice, Ann Gravells is considered to be an expert.

Teaching and learning (anngravells.com)

Facilitation

Feedback/Review

Part Five

Chapter 15—Leadership and Management

Part Six

Chapter 16—RPAS Operational Oversight (Courtesy of heliguy)

heliguy

Unit 9, Jupiter Court, Orion Business Park, North Shields NE29 7SE · 12 mi
0191 296 1024
www.heliguy.com/

Insurance—EC Regulation 785/2004

https://eur-lex.europa.eu/legal-content/PT/TXT/?uri=CELEX:32004R0785

CAA's Skywise Service

http://skywise.caa.co.uk/

UK Regulation (EU) 2016/679

REGULATION (EU) 2016/679 OF THE EUROPEAN PARLIAMENT AND OF THE COUNCIL—of 27 April 2016—on the protection of natural persons with regard to the processing of personal data and on the free movement of such data, and repealing Directive 95/46/EC (General Data Protection Regulation) (europa.eu)

Data Protection Act (DPA 2018) and the General Data Protection Regulation (GDPR 2018)

Data Protection Act 2018 (legislation.gov.uk)

CAP2378 Consultation Document for the Acceptable Means of Compliance and Guidance Material to Regulation (EU) 2019/947 as retained (and amended in UK domestic law) Under the European Union (withdrawal) Act 2018.

Acceptable Means of Compliance and Guidance Material to UK Regulation (EU) 2019/947—Civil Aviation Authority—Citizen Space (caa.co.uk)

Chapter 17—Understanding the Importance of UAS Maintenance (Courtesy of heliguy)

Health and Safety Executive (HSE)

The Provision and Use of Work Equipment Regulations 1998 (PUWER). www.hse.gov.uk/work-equipment-machinery/puwer.htm

Chapter 18—Organising a Flight Safety Meeting (Courtesy of heliguy)

Occurrence reporting—UK Regulation 376/2014 and implementing regulation 1018/2015.

Occurrence Reporting | Civil Aviation Authority (caa.co.uk)

UK and European Reporting link

https://aviationreporting.eu/

Chapter 19—Flight Readiness Review (courtesy of heliguy)

Chapter 20—Future of Unmanned Aircraft Systems (courtesy of heliguy)

Chapter 21—Preparing an Operational Safety Case (courtesy of heliguy)

Chapter 22—Understanding the Auditing Process for UAS Operators (courtesy of heliguy)

www.heliguy.com/products/drone-osc-consultation

GENERAL INTEREST LINKS

Civil Aviation Authority—Australian Government (CASA)—Av safety podcasts and videos

www.casa.gov.au/

European Helicopter Safety Team (EHST)

www.easa.europa.eu/community/topics/ehest-heritage-useful-sms-tools

HE 4: Decision-Making

HE 5: Risk Management in Training

HE 8: Threat Error Management

HE 9: Automation and Flight Path Management

SKYBRARY AVIATION SAFETY—This is one of the best sites for aviation information

https://skybrary.aero/

Transport Canada

https://tc.canada.ca/en/aviation

DRONE MANUFACTURERS

Top drone manufacturers of 2023

1. **DJI (China) continues to dominate civil drones**—www.dji.com/uk
2. **SKYDIO (USA)**—www.skydio.com/
3. **XAG (China)**—www.xa.com/en
4. **PARROT (France)**—www.parrot.com/en
5. **JOUAV (China)**—www.jouav.com/

Dual-Use (Civil and Government) Drones

1. **Insitu (USA)**—www.insitu.com/
2. **Schiebel (Austria)**—https://schiebel.net/
3. **Edge Autonomy (USA)**—https://edgeautonomy.io/
4. **Quantum Systems (Germany)**—www.quantum-systems.com/
5. **ideaForge (India)**—www.ideaforge.co.in/?trk=public_post_main-feed-card_reshare-text

VIDEO LINKS

KLM Tenerife—https://youtu.be/8K9fUc5O_G0

US Airways G01549 A320 Hudson River—https://youtu.be/mwkdmMTCCPg

Trans Asia 235—https://youtu.be/I8l1_ZDzxcM

Avianca 52—https://youtu.be/0Ql1Bjm4Ptk

United Airlines 232—https://youtu.be/fG-6nHwfyts

Saudia 163—https://youtu.be/emq9EoIHCX8

Dan Air DA 1008—https://youtu.be/LPoUWpf-CwY

Teterboro Lear Jet—https://youtu.be/2dE6LROPK58

German Wings 9525—https://youtu.be/36lg28M3Hb0

Air France 447—https://youtu.be/kERSSRJant0?t=4

John and the duck—https://youtu.be/VaCZ0Ey0Imw

REGULATORY PUBLICATIONS

FAA—www.faa.gov/

FLIGHT SAFETY FOUNDATION—https://flightsafety.org/

TRANSPORT CANADA—https://tc.canada.ca/en/aviation

UK CAA

CAP 722—Unmanned Aircraft System Operations in UK Airspace—Unmanned Aircraft System Operations in UK Airspace Guidance
www.caa.co.uk/our-work/publications/documents/content/cap-7

CAP 722A: Unmanned Aircraft System Operations in UK Airspace—Unmanned Aircraft System Operations in UK Airspace Unmanned Aircraft System Operations in UK AirspaceUnmanned Aircraft System Operations in UK Airspace—Unmanned Aircraft System Operations in UK Airspace Operating Safety Cases
www.caa.co.uk/our-work/publications/documents/content/cap-722a/

CAP 716: Aviation Maintenance Human Factors (EASA Part-145) | Civil Aviation Authority (caa.co.uk)

CAP 737—Flight Crew Human Factors Handbook
www.caa.co.uk/our-work/publications/documents/content/cap-737/

CAP 726 Auditing—www.caa.co.uk/publication/CAP726
CAP 795—www.caa.co.uk/our-work/publications/documents/content/cap-795/

CAP 1496—New Aviation Reporting Portal

CAP1496: New Aviation Reporting Portal | Civil Aviation Authority (caa.co.uk)

CAP 1607—www.caa.co.uk/our-work/publications/documents/content/cap1607/

CAP 382 Occurrence Reporting—https://skybrary.aero/sites/default/files/bookshelf/545.pdf

CAP 393 Air Navigation Order 2016 www.caa.co.uk/publication/download/12231

CAP 2378—https://consultations.caa.co.uk/corporate-communications/amc-and-gm-to-uk-regulation-eu-2019–947/

Index

E

F

G

H

I

J

K

L

Printed in the United States
by Baker & Taylor Publisher Services